DESIGN AND PERFORMANCE OF 3G WIRELESS NETWORKS AND WIRELESS LANS

T0226023

DESIGN AND PERFORMANCE OF 3G WIRELESS NETWORKS AND WIRELESS LANS

MOOI CHOO CHUAH
Lehigh University

QINQING ZHANG
Bell Laboratories, Lucent Technologies

 Springer

Mooi Choo Chuah
Lehigh University
USA

Qinqing Zhang
Bell Laboratories, Lucent Technologies
USA

Design and Performance of 3G Wireless Networks and Wireless LANs

ISBN 978-1-4419-3696-7

e-ISBN 0-387-24153-1
e-ISBN 978-0-387-24153-1

Printed on acid-free paper.

Printed in the United States of America.

9 8 7 6 5 4 3 2 1

springeronline.com

This book is dedicated to our families.

Contents

Preface

Cellular phones, especially those enabled by second-generation telecommunication systems, have had tremendous impacts on our daily lives. In some countries such as India, the number of cellular phone subscribers has far exceeded the number of wired phone subscribers. Meanwhile, the Internet has also significantly changed our daily lives. More and more e-commerce applications have been introduced while the number of Internet users has skyrocketed over the recent five years. The mobile workforce has also tremendously increased in size. Mobile workers expect to be able to use the Internet while on the move. However, the data handling capabilities of second-generation systems are limited. Thus, third-generation (3G) cellular systems such as UMTS (Universal Mobile Telecommunication Systems) and CDMA2000 (Code-Division Multiple Access) Systems are designed to provide high bit rate data services that enable multimedia communications. Such third-generation cellular systems allow high-quality images and video to be transmitted and received. The third-generation cellular systems also provide open-access capabilities where value-added services, e.g., location-based services, can be introduced by third-party providers. While the 3G standards are being drafted, and equipment for third-generation cellular systems is being designed, wireless LAN systems are introduced into our daily lives to meet our demand for wireless data services while on the move. This book describes the network architectures of UMTS and CDMA2000 systems and how major network elements within the 3G networks can be designed. In addition, this book provides discussions on how the end-to-end performance for voice and data services can be determined. It also provides guidelines on how the radio access networks and core networks can be engineered. Last but not least, this book describes the various wireless LAN standards and how voice and data services can be offered in wireless LAN systems.

The book is organized as follows: Chapter 1 provides an introduction to wireless communication concepts. It briefly discusses the first- and second-generation systems that are based on Frequency Division Multiple Access (FDMA) and Time Division Multiple Access (TDMA) technologies, and the spread spectrum-based communication systems. Then, it briefly discusses common techniques used in spread-spectrum communications, e.g., power control, soft handoff, adaptive modulation and coding, and multiuser diversity. Chapter 2 provides an introduction to wireless systems. It discusses generic wireless system architecture and how the system operates, e.g., the registration of mobile phones, how mobile phones initiate calls, how calls are delivered, what happens when mobile phone users move, and how intra/inter-system handoffs are carried out. Chapter 3 provides an introduction to traffic engineering issues. Service providers are interested in maximizing their revenue via offerings of high-value services while maintaining high utilization of their installed infrastructure. Thus, traffic engineering is required since different applications have different quality of service requirements. Traffic models for different applications need to be developed. Chapter 3 discusses techniques that one can use to determine the traffic models for different applications, e.g., WWW-browsing and emails. It also discusses the different parameters used to describe circuit-switched and packet-switched services. Chapter 4 describes the network architectures for UMTS and CDMA2000 systems. Chapter 5 analyzes the airlink interface capacity and performance for UMTS/CDMA2000 systems. Chapter 6 describes how the 3G base station can be designed to meet certain performance requirements. Chapter 7 describes how the 3G base station controller can be designed and how the radio access networks can be engineered. Techniques that can be used to reduce the OPEX of the radio access networks are also discussed. Chapter 8 describes how the core network elements can be designed. Chapter 9 describes the end-to-end performance of voice and data services in 3G systems. Chapter 10 provides a high-level description of the various 802.11-based wireless LAN systems. Chapter 11 describes the medium access control (MAC) and quality of service (QoS) features in 802.11-based wireless LAN systems. Chapter 12 discusses the upcoming 3G features.

This book is aimed at operators, network manufacturers, service providers, engineers, university students, and academicians who are interested in understanding how 3G and wireless LAN systems should be designed and engineered.

Mooi Choo Chuah
Qinqing Zhang

Acknowledgments

The authors would like to acknowledge many colleagues who are or were from Bell Laboratories, Lucent Technologies for their contributions to the research work done with the authors that are reported in this book. The authors would like to thank the anonymous reviewers and Dr. D. Wong from Malaysian University of Science and Technology for providing useful suggestions to improve the content and presentations in the book.

The authors would also like to thank Springer's supporting staff members for answering numerous questions during the book writing process.

We are extremely grateful to our families for their patience and support, especially during the late night and weekend writing sessions.

Special thanks are due to our employers, Lucent Technologies and Lehigh University, for supporting and encouraging such an effort. Specifically, the authors would like to thank Dr. Victor B. Lawrence, the former Vice President of Advanced Communications Technologies, for his support and encouragement during the initial phase of our book writing process. Special thanks are due to Lucent Technologies, IEEE, 3GPP for giving us permission to use diagrams and illustrations for which they own the copyrights.

The authors welcome any comments and suggestions for improvements or changes that could be implemented in forthcoming editions of this book. The email address for gathering such information is 3gbook@cse.lehigh.edu.

Mooi Choo Chuah
Qinqing Zhang

Author Biographies

Mooi Choo Chuah is currently an associate professor at Lehigh University. She received her B. Eng. with Honors from the University of Malaya, and MS and Ph.D. degrees in electrical engineering from the University of California, San Diego. She joined Bell Laboratories, Holmdel, New Jersey in 1991. She was promoted to be Distinguished Member of Technical Staff in 1999 and was made a technical manager in 2001. While at Bell Laboratories, she worked on wireless communications, IP/MPLS protocol designs, and has been a key technical contributor to various business units and product teams at Lucent. She has been awarded 34 patents and has 25 more pending. Her current research interests include heterogeneous network system and protocol design, network/computer system security, disruption tolerant networking, and ad-hoc/sensor network design.

Qinqing Zhang is a Member of Technical Staff at Bell Labs, Lucent Technologies. She received her B.S. and M.S.E. degrees in Electronics Engineering from Tsinghua University, Beijing, China, M.S. and Ph.D. degrees in Electrical Engineering from the University of Pennsylvania, Philadelphia. Since joining Bell Labs in 1998, she has been working on the design and performance analysis of wireline and wireless communication systems and networks, radio resource management, algorithms and protocol designs, and traffic engineering. She has been awarded 6 patents and has 14 patent applications pending. She is an adjunct assistant professor at the Unversity of Pennsylvania. She is a senior member of IEEE. She serves on the editorial board of IEEE Transactions on Wireless Communications and technical program committees of various IEEE conferences.

Chapter 1

INTRODUCTION TO WIRELESS COMMUNICATIONS

1. INTRODUCTION

The birth of wireless communications dates from the late 1800s, when M.G. Marconi did the pioneer work establishing the first successful radio link between a land-based station and a tugboat. Since then, wireless communication systems have been developing and evolving with a furious pace. The number of mobile subscribers has been growing tremendously in the past decades. The number of mobile subscribers throughout the world increased from just a few thousand in the early 20th century to close to 1.5 billion in 2004.

The early wireless systems consisted of a base station with a high-power transmitter and served a large geographic area. Each base station could serve only a small number of users and was costly as well. The systems were isolated from each other and only a few of them communicated with the public switched telephone networks. Today, the cellular systems consist of a cluster of base stations with low-power radio transmitters. Each base station serves a small cell within a large geographic area. The total number of users served is increased because of channel reuse and also larger frequency bandwidth. The cellular systems connect with each other via mobile switching and directly access the public switched telephone networks. The most advertised advantage of wireless communication systems is that a mobile user can make a phone call *anywhere* and *anytime*.

1.1 Technology Evolution

In the early stages, wireless communication systems were dominated by military usage and supported according to military needs and requirements. During the last half a century, with increasing civil applications of mobile services, commercial wireless communication systems have been taking the lead.

1.1.1   Basic Principles

In a cellular network, an entire geographic area is divided into cells, with each cell being served by a base station. Because of the low transmission power at the base station, the same channels can be reused again in another cell without causing too much interference. The configuration and planning of the cell is chosen to minimize the interference from another cell and thus maximum capacity can be achieved. The cell is usually depicted as a hexagon, but in reality the actual shape varies according to the geographic environment and radio propagation. Channel allocation is chosen based on the density of the users. If a cell has many users to serve, usually more channels are allocated. The channels are then reused in adjacent cells or cluster of cells. The spatial separation of the cells with the same radio channels, in conjunction with the low transmission power and antenna orientation, keeps the co-channel interference at an acceptable level.

Mobility is one of the key features in wireless communication systems. There is a need to track the users moving into different cells and changing radio channels. A mobile switched to another channel in a different cell is called handoff. A signaling and call processing procedure is needed to support user mobility and handoff such that a mobile phone can be completed successfully. Paging is another key feature in cellular systems. It uses a common shared channel to locate the users within the service area and to broadcast some signaling messages.

1.1.2   Multiple Access Technique

Multiple access is a technique to allow users to share a communication medium so that the overall capacity can be increased. There are three commonly used multiple access schemes: Frequency Division Multiple Access (FDMA), Time Division Multiple Access (TDMA) and Code Division Multiple Access (CDMA).

In FDMA, each call is assigned its own band of frequency for the duration of the call. The entire frequency band is divided into many small individual channels for users to access. In TDMA, users share the same band of frequencies. Each call is assigned a different time slot for its transmission. In CDMA, users share the same band of frequencies and time slots. Each call is assigned a unique code, which can spread the spectrum to the entire frequency band. The spectrum spread calls are sent on top of each other simultaneously, and are separated at the receiver by an inverse operation of the unique codes. A combination of the three multiple access schemes can also be applied.

1.1.3   System Implementations

We describe briefly the popular specific implementations of wireless communication systems.

1.1.3.1   Advanced Mobile Phone Service (AMPS)

The Advanced Mobile Phone Service (AMPS) was the very first implementation of the cellular mobile systems. It is an analog system in which each user fully occupies the radio channel of 30 KHz.

Each base station in AMPS operates in the 800–900 MHz band. It utilizes the frequency division duplex (FDD) in which the uplink and downlink transmission is carried at different frequencies. Each carrier has 416 two-way radio channels divided into a cluster of seven cells. Each cell can support about 60 channels on average.

The analog AMPS system was later evolved to a digital system (DAMPS), also known as IS-54. In DAMPS, digital coding together with the TDMA technique is used to allow three users in the 30-KHz radio channel. The capacity is thus greatly increased.

1.1.3.2   Global System for Mobile (GSM) Communications

The Global System for Mobile (GSM) communications was introduced in 1992 as a European standard and has achieved much worldwide success.

The GSM system operates in the 800-MHz band and 1800-MHz band in Europe. The 1900-MHz band system is intended for the United States. It uses FDD for uplink and downlink transmission. Each radio channel has 200-KHz bandwidth. The GSM900 has total of 124 two-way channels assigned to a cluster of seven cells, while the GSM1800 has 374 two-way channels.

The multiple-access technique in GSM is TDMA. Eight users share each 200-KHz channel. Equivalently each user has 25-KHz bandwidth for use, which is comparable to the bandwidth assigned to AMPS users. The speech coding and compression in GSM is called the regular pulse-excited, long-term prediction (RPE-LTP) and is also known as residual-excited linear prediction (RELT). The coded bit rate is 13 Kbps.

Error correction and interleaving is introduced in the GSM system to combat the channel errors. The modulation scheme is called Gaussian minimum shift keying (GMSK), which is one type of frequency shift keying (FSK) technique.

Slow frequency hopping (SFH) is used at a slow frame rate. Each frame is sent in a repetitive pattern, hopping from one frequency to another through all available channels. Frequency hopping reduces the effect of fading and thus improves the link performance.

Mobile-assisted handoff is performed in the GSM systems. The mobile monitors the received signal strength and quality from different cells and sends back a report periodically. Based on the report, the base station decides when to switch the mobile to another channel.

1.1.3.3   General Packet Radio Service (GPRS) Systems

The general packet radio service (GPRS) is an enhancement to the GSM mobile communication systems that support packet data. It has been standardized by ETSI, the European Telecommunication Standards Institute [GPRS1][GPRS2].

GPRS uses a packet switching to transmit high-speed data and signaling more efficiently than the GSM systems. It optimizes the network and radio resource usage. It maintains strict separation of the radio subsystem and network subsystem, allowing the network subsystem to be used with other radio access technologies.

GPRS defines new radio channels and allows dynamic channel allocation for each user. One and up to eight time slots per TDMA frame can be assigned to an active user. Various channel coding schemes are defined to allow bit rates from 9 Kbps to more than 150 Kbps per user.

GPRS supports internetworking with IP and X.25 networks. Applications based on the standard protocols can be transferred over the GPRS radio channels. New network nodes are introduced in the GPRS core network to facilitate the security, internetworking, and mobility management.

GPRS is designed to support intermittent and bursty data transmission. Four different quality of service classes are defined. User data are transferred transparently between the mobile station and the external data networks via encapsulation and tunneling. User data can be compressed and protected with retransmission for efficiency and reliability.

1.1.3.4   Enhanced Data Rates for Global Evolution (EDGE)

The enhanced data rates for global evolution (EDGE) is the new radio interface technology to boost network capacity and user data rates for GSM/GPRS networks [Zan98]. It has been standardized by ETSI and also in the United States as part of the IS-136 standards.

EDGE gives incumbent GSM operators the opportunity to offer data services at speeds that are close to those available on the third-generation wireless networks (which will be described in more details in later chapters.) It increases the GSM/GPRS data rates by up to three times. EDGE enables services such as emails, multimedia services, Web browsing, and video conferencing to be easily accessible from a mobile terminal.

EDGE uses the same TDMA frame structure, logic channel, and 200-KHz channel bandwidth as the GSM networks. It introduces the 8-PSK modulation and can provide data throughput over 400 Kbps per carrier. It supports peak rates up to 473 Kbps per user. Adaptive modulation and coding scheme is applied to the EDGE system to increase the system efficiency.

A key design feature in EDGE systems is the link quality control, through link adaptation and incremental redundancy. A link adaptation technique regularly estimates the channel quality and subsequently selects the most appropriate modulation and coding scheme for the new transmission to maximize the user bit rate. In the incremental redundancy scheme, information is first sent with very little coding, yielding a high bit rate if decoding is successful. If the decoding fails, more coding bits are sent and generate a low bit rate.

EDGE devices are backwards compatible with GPRS and will be able to operate on GPRS networks where EDGE has not been deployed.

1.1.3.5 Spread Spectrum Communication

Spread spectrum communication technology uses a communication bandwidth much larger than the information bandwidth. The signal bit stream is coded and spread over the entire spectrum space using a unique signature code. The receiver searches the unique signature code and separates the desired signal from others. This technique is called CDMA [Lee91].

Another spread spectrum technique is frequency hopping, in which each signal stream switches frequency channel in a repetitive pattern. The receiver searches the appropriate pattern for the desired signal. As discussed earlier, slow frequency hopping is used in the GSM systems.

Spread spectrum technique has several advantages over the traditional communication schemes. First, it suppresses the intentional or unintentional interference by an amount proportional to the spreading factor. Therefore spread spectrum communication is less prone to interference. Second, it increases the accuracy of position location and velocity estimation in proportion to the spreading factor. Third, the spread signal has low detection probability by an unknown device and thus the security of the transmission is improved. Finally, it allows more users to access the same spectrum space and increases the system capacity.

Spread spectrum technology has been used in military communication for over half a century because of its unique advantages over other technologies. The CDMA spread spectrum has been advocated and developed for commercial cellular systems by Qualcomm, Inc. The spread spectrum system was formalized by North America and then adopted by the Cellular Telephone Industry Association (CTIA) as the IS-95 standard. It is also known as CDMA-One. It operates in the same 900-MHz frequency band as AMPS. Each radio channel has a 1.25-MHz bandwidth. The new personal communication system (PCS) operates in the 1900-MHz band.

The speech coding and compression in IS-95 and CDMA2000 systems is called Qualcomm code-excited linear prediction (QCELP). The coded bit rate varies adaptively from 1 Kbps to 8 Kbps. The speech bits together with the error correction codes result in a gross bit rate varying from 2.4 Kbps to 19.2 Kbps. The bit stream is multiplied by a pseudorandom code, which is called the spreading code. The multiplication of the spreading code has the effect of spreading the bit stream to a much greater bandwidth. At the receiver, the appropriate pseudorandom code is applied to extract the desired signal. The other undesired signals appear as random noise and are suppressed and ignored.

CDMA spread spectrum has become the dominating technology in the third generation (3G) wireless communication standards. It has been adopted by both CDMA2000 and Universal Mobile Telecommunication System (UMTS) standards.

In this book, we describe in detail traffic engineering design issues in the 3G CDMA systems.

1.2   Techniques in Wireless Communications

1.2.1   Power Control

Power control is one of the most important design features in wireless communication including FDMA, TDMA, and CDMA systems [Nov2000]. It ensures each user transmits and receives at a proper energy level to convey information successfully while reducing the interference to other users.

Power control is needed in FDMA and TDMA systems because of the co-channel interference management. This type of interference is caused by the frequency reuse in the limited available spectrum. Via a proper power level adjustment, the co-channel interference can be reduced. This allows a higher frequency reuse factor and thus increases the system capacity.

Power control is the most essential requirement in CDMA systems [Zen93][Gra95][Han99]. Without power control, all the mobiles transmit to the base station with the same power not taking into account path loss and fading effect. Mobiles close to the base station will cause significant interference to mobiles that are farther away from the base station. This effect is the so-called near/far effect. Therefore, a well-designed power control algorithm is crucial for proper operation of a CDMA system. In the absence of power control, the system capacity is very low compared to other systems.

Another advantage of power control is that it can prolong battery life by using a minimum required transmission power.

Power control on a reverse link is more stringent than on a forward link because of the near/far effect. On a forward link, power control is still necessary to reduce the inter-cell interference.

Power control can be operated in a centralized form or a distributed form. A centralized controller obtains the information of all the established connections and channel gains, and controls the transmission power level. The centralized approach can optimize the power usage of the entire or part of the network and thus is very efficient. It requires extensive control signaling in the network, however, and is difficult to apply in practice.

The distributed controller controls only one transmitter of a single connection. It controls transmission power based on local information such as the signal-to-interference ratio and channel gains of the specific connection. It is easy to implement and thus is widely used in actual systems.

Power control techniques can be categorized into two classes: closed-loop power control and open-loop power control [Cho98]. In closed-loop power control, based on the measurement of the link quality, the base station sends a power control command instructing the mobile to increase or decrease its transmission power level. In open-loop power control, the mobile adjusts its transmission power based on the received signaling power from the base station. Since the propagation loss is not symmetric, the open-loop power control may not be effective. Thus a closed-loop power control must be in place to manage the power level.

The closed-loop power control is feasible in a terrestrial cellular environment. The open-loop power control is more appropriate for satellite communications where the round trip propagation delay is too large for the closed-loop power control to track the fading variation.

The closed-loop power control consists of two parts: an inner loop and an outer loop that are operated concurrently. The inner loop is based on the measurement, for example, signal-to-interference ratio (SIR). The receiver estimates the received SIR and compares it to a target value. If the received SIR is lower than the target SIR, the receiver commands the transmitter to increase its power. If the received SIR is higher than the target, the receiver commands the transmitter to decrease its power. The outer loop is based on the link quality, typically the frame error rate (FER) or bit error rate (BER). The receiver estimates the FER or BER and adjusts the target SIR accordingly. The outer loop power control is especially important when the channel state changes over time. A pure SIR-based control cannot guarantee a certain link performance. Therefore outer loop power control is essential in maintaining a user's link quality.

There has been great effort in designing power control algorithms for the CDMA systems. The combination of power control with multiuser detection [Ulu98] and beam forming techniques [Ras98] is very promising in improving the spectrum efficiency in CDMA systems.

1.2.2   Soft Handoff

Soft handoff is a unique feature in CDMA systems. It is a smooth transition of a phone transferred from one cell to another cell. In CDMA systems, since all the cells operate at the same frequency, it makes it possible for a user to send the same call simultaneously to two or more base stations. On the contrary, in an FDMA/TDMA system, a given slot on a given frequency channel cannot be reused by adjacent cells. When a user moves from one cell to another, it needs to switch its channel and frequency all at once, which is the so-called hard handoff.

Soft handoff can offer superior performance improvement compared to hard handoff in terms of the reverse link capacity. The capacity increase comes from the macro diversity gain from the soft handoff. Signals of the same call arrive at multiple base stations through different paths. A controller can choose the signal from the best path and decode it successfully. On the forward link, multiple base stations can send the signals to the same mobile. The mobile can combine the received signals from the different base stations and improve the performance.

Soft handoff provides a smooth and more reliable handoff between base stations when a mobile moves from one cell to another cell. In a heavily loaded system, with soft handoff and proper power control, the system capacity can be doubled. In a lightly loaded system, the cell coverage can be doubled because of soft handoff.

The soft handoff process consists of multiple steps. First, the mobile monitors the received signal strength from different cells. When it detects a strong signal from a base station, it informs the system and requests to add the cell to its active sets. The communication link between the mobile and the cell is called a leg. After the system adds a leg to the mobile's active set, the mobile starts transmitting to both cells. As the mobile continues moving, the signal from the first cell fades away. The mobile informs the system and drops the leg eventually. The adding and dropping of legs may occur several times depending on the mobile speed and propagation environment.

The performance of soft handoff is very sensitive to the settings of the parameters in the actual implementation. The parameters can be optimized to achieve the best trade-off between performance enhancement and implementation complexity.

1.2.3 Adaptive Modulation and Coding

Adaptive modulation and coding (AMC) has been widely used to match the transmission parameters to the time varying channels. It greatly improves the spectrum efficiency and system performance.

Because the fading channel is time varying and error prone, static configuration of the modulation and coding scheme has to be designed conservatively to maintain the required link quality and performance, and results in a low efficient use of the radio resource. In adaptive modulation and coding schemes, the channel quality is measured and estimated regularly. Based on the channel state, a proper modulation and coding scheme is chosen for the upcoming transmission so that the user bit rate can be maximized.

To make effective use of AMC, reliable channel quality information is essential. Various techniques have been explored for channel estimation and predication based on the measurement data.

Adaptive modulation and coding has been incorporated in the new wireless communication systems. In EDGE, link adaptation is used to increase the user bit rate and maximize the system throughput. In 3G wireless systems including both CDMA2000 and UMTS, adaptive modulation and coding has been used to provide high-speed data transmission.

1.2.4   Space–Time Coding and Multiuser Diversity

Space–time coding (STC) was first introduced [Tar98] to provide transmission diversity for multiple-antenna fading channels. It is a design of combining coding, modulation, transmission and receive diversity schemes [Tar98][Tar99_1][Tar99_2].

Space–time coding offers an effective transmission-antenna diversity technique to combat fading. It is a highly bandwidth efficient approach that takes advantage of the spatial dimension by transmitting a number of data streams using multiple antennas.

There are various approaches to the coding structures, including space–time trellis coded modulation, space–time turbo codes and also space–time layered structure. The essential issue in designing space–time coding structure is to take advantage of the multipath effects to achieve very high spectrum efficiency.

There are two main types of STCs: space–time block codes (STBC) and space–time trellis codes (STTC) [Ala98][San2001]. Space–time block codes contain a block of input symbol, generating a matrix whose columns represent time and rows represent antennas. They provide full diversity with a simple decoding scheme. Space–time trellis codes operate on one input symbol at a time and produce a sequence of symbols. The length of the vector symbols represents antennas. In addition to the full diversity, the space–time trellis codes can provide coding gain as well. But they are very difficult to design and require much more complex encoder and decoder than the space–time block codes.

The space–time coding technique can increase the capacity by an order of magnitude. It has been studied intensively in both academia and industry and has been adopted in the third-generation wireless communication systems.

1.3   Summary

Wireless communication systems have experienced tremendous growth in the past century. The commercial cellular systems evolved from the analog system to the digital system rapidly. Wireless technology has progressed through the first-generation (1G), the second-generation (2G), and the current third-generation (3G) systems. The services in wireless systems have expanded from voice only to today's high-speed data, multimedia applications and wireless Internet. The key techniques in wireless communications have been exploited and the technology revolution continues its development.

1.4   References

[Ala98]   S. M. Alamouti, "Simple Transmit Diversity Technique for Wireless Communications," *IEEE Journal on Select Areas in Communications*, vol. 16, pp. 1451–1458, 1998.

[Cho98]   A. Chockalingam and L. B. Milstein, "Open Loop Power Control Performance in DS-CDMA Networks with Frequency Selective Fading and Non-Stationary Base Stations," *Wireless Networks 4*, 1998, pp. 249–261.

[GPRS1]   ETSI TS 03 64 V5.1.0, "Digital Cellular Telecommunications System (Phase 2+); General Packet Radio Service (GPRS); Overall Description of the GPRS Radio Interface; Stage 2 (GSM 03.64, v.5.1.0)." November 1997.

[GPRS2]   ETSI GSM 02.60, "General Packet Radio Service (GPRS); Service Description; Stage 1," v.7.0.0, April 1998.

[Gra95]   S. A. Grandhi, J. Zander, and R. Yates, "Constrained Power Control," *Wireless Personal Communications*, vol. 1, No. 4, 1995.

[Han99]   S. V. Hanly and D. Tse, "Power Control and Capacity of Spread-Spectrum Wireless Networks," *Automatica*, vol. 35, no. 12, Dec. 1999, pp. 1987–2012.

[Lee91]   W. C. Y. Lee, "Overview of Cellular CDMA," *IEEE Transaction on Vehicular Technology*, vol. 40, no. 2, May 1991.

[Nov2000]   D. M. Novakovic and M. L. Dukic, "Evolution of the Power Control Techniques for DS-CDMA Toward 3G Wireless Communication Systems," *IEEE Communication Surveys & Tutorials*, 4th quarter issue, 2000.

[Ras98]   F. Rashid-Farrokhi, L. Tassiulas, and K. J. R. Liu, "Joint Optimal Power Control and Beamforming in Wireless Networks Using Antenna Arrays," *IEEE Transaction on Communications*, vol. 46, no. 10, Oct. 1998.

[San2001]   S. Sandhu and A. J. Paulraj, "Space-Time Block Codes versus Space-Time Trellis Codes," *Proceedings of ICC2001*.

[Tar98]   V. Tarokh, N. Seshadri, and A.R. Calderbank, "Space-Time Codes for High Data Rates Wireless Communications: Performance Criterion and Code Construction," *IEEE Transaction on Information Theory*, vol. 44, pp. 744–765, 1998.

[Tar99_1]   V. Tarokh, H. Jafarkhani, and A.R. Calderbank, "Space-Time Block Coding from Orthogonal Designs," *IEEE Transaction on Information Theory*, vol. 45, pp. 1456–1467, 1999.

[Tar99_2]   V. Tarokh, H. Jafarkhani, and A.R. Calderbank, "Space-Time Block Coding for Wireless Communications: Performance Results," IEEE Journal on Select Areas in Communications, vol. 17, pp. 451–460, 1999.

[Ulu98]   S. Ulukus and R. D. Yates, "Adaptive Power Control and MMSE Interference Suppression," *Baltzer/ACM Wireless Networks*, vol. 4, no. 6, June 1998, pp. 489–496.

[Zan98] K. Zangi, A. Furuskar and M. Hook, "EDGE: Enhanced Data Rates for Global Evolution of GSM and IS-136," Proceedings of Multi Dimensional Mobile Communications, 1998.

[Zen93] J. Zender, "Transmitter Power Control for Co-Channel Interference Management in Cellular Radio Systems," *Proceedings of 4th WINLAB Workshop*, New Brunswick, New Jersey, USA, Oct. 1993.

Chapter 2

INTRODUCTION TO WIRELESS SYSTEMS

2. INTRODUCTION

The main goal of a wireless system is to provide information services, e.g., voice and data to mobile users via wireless medium. Since the mid-1990s, the number of cellular telephone subscribers worldwide has grown rapidly. The widespread success of cellular has also led to the development of newer wireless systems. In this chapter, in Section 2.2. we will first describe the basic components of a generic wireless system architecture that provides voice services. Next, in Section 2.3, we describe how information can be transported in wireless networks. Then, in Section 2.4, we describe the first- and second-generation cellular radio networks. In Section 2.5, we discuss the deficiencies of first- and second-generation wireless systems that prevent them from offering efficient wireless data services. We then describe in Section 2.6, the Cellular Digital Packet Data (CDPD) network – an overlay network over an existing second-generation wireless systems to provide wireless data services. Research continues in all areas, e.g., wireless system architecture, service platforms, airlink enhancements, etc. to provide a better system that can overcome the limitations of the current systems and meet new emerging requirements, e.g., higher airlink bandwidths, ease of providing new services, and support of more sophisticated services such as wireless auctions, etc. Third-generation networks are designed with the intention to meet these challenges. We briefly discuss third-generation networks in Section 2.7 and defer detailed discussion to subsequent chapters. In Section 2.8, we describe some of the transport choices for wireless backhaul networks. The transport choices depend on several factors, e.g., whether the carriers are Greenfield operators or incumbents, the availability of certain IP-based features in relevant network elements, e.g., base stations and radio network controllers, Quality of Service (QoS) requirement, etc. In Section 2.9, we present some examples of the end-to-end protocol stacks for supporting circuit-switched and packet-switched services. Lastly, we describe briefly one important function in wireless networks, i.e., RLC

(Radio Link Control) and MAC (Medium Access Control) function, and how it affects the end-to-end performance.

2.1   Generic Wireless System Architecture

A wireless system that provides voice services consists of the following network elements:

- Mobile Stations
  To use the voice services, a mobile user needs to have a cellular phone, which can generate radio signals carrying both signaling messages and voice traffic.   The radio channels carrying signaling messages are referred to as control channels while the radio channels carrying voice traffic are referred to as traffic channels. Some in-band signaling messages are carried within the traffic channels as well.

- Base Station
  Cellular phones must be able to communicate with a base station via the radio channels. A base station is a collection of equipment that communicates by radios with many cellular phones within a certain geographic area whereby the radio signals sent by the phones can be received correctly at the base station and similarly the phones can receive the radio signals sent by the base station correctly. Such a geographic area is referred to as a cell. The base station terminates the radio signals, extracts the voice traffic, and packages the voice traffic into forms appropriate to be sent to a controller that determines how the voice traffic needs to be routed.   Such a controller is often referred to as a Mobile Switching Center (MSC), which is described next.

- Mobile Switching Center (MSC)
  A cellular system consists of typically thousands of cells, each with its own base station. The base stations are connected via wired trunks to a mobile switching center where the phone calls are switched to appropriate destinations. For a service provider's network that covers a wide geographic area, there may be more than one MSC and the MSCs will be connected via wired trunks.

- Home/Visiting Databases

Mobile users subscribe to voice services from a wireless service provider (WSP) and pay a monthly fee to enjoy the benefit of being able to place phone calls anywhere they want (within the coverage area of the WSP). Thus, the WSP needs to provide a database with the subscriber's personal information, e.g., the service plan he/she subscribes to, the unique phone number assigned to the subscriber, and the current location of the subscriber whenever he/she powers up the phone. Such a database is often referred to as the home location register (HLR) if it is located within the part of the WSP's network where the subscriber buys the service. The mobile users may roam to a different site/state where the WSP still offers voice services. Such users are required to register with a local MSC and a database containing visiting users' information that is often referred to as the Visiting Location Register (VLR). An authentication center is often provided to match the Mobile Identification Number (MIN) and Electronic Serial Number (ESN) of every active cellular phone in the system with the information stored in the home location register. If a phone does not match the data in the home database, the authentication center instructs the MSC to disable the questionable phone, thereby preventing such phones from using the network.

Several important activities take place in a wireless system: mobile registrations, call initiations, call delivery, mobility management of mobile stations and intra/inter-system handoffs of mobile stations. We describe each of these activities next.

2.1.1   Registration and Call Initiation

When a mobile user's phone is activated, it scans for the strongest control channel in its vicinity. Each control channel carries signals from one base station in the vicinity of the cellular phone. Then, the cellular phone tunes to that strongest control channel and decodes the signals in the channel. The phone acquires several important pieces of system information, e.g., the system identifier, how much power it needs to transmit to send information to the base station, which radio channel it needs to use to transmit such information, etc. The phone registers itself with the network. When a user initiates a phone call, the user keys in the phone number and hits the SEND button. The phone then initiates a service request procedure with the cellular network. On receiving such a request, the base station relays the information to the MSC to which it is connected. The MSC analyzes the dialed digits and communicates with PSTN or other MSC to route the call to its destination.

2.1.2   Mobility Management

Now, other mobile users or fixed wired line phone users may desire to place a call to this mobile user. The PSTN consults the home location register of the user to determine the MSC where the mobile user is registered currently. Then, the called signals are relayed to that MSC. The MSC will then initiate a paging procedure to inform the mobile user of an incoming call.

Thus, it is important for a mobile user to inform the network of its whereabouts so that calls may be routed to such a roamer as it moves through the coverage area of different MSCs. When a user roams into a new area covered by the same or different service provider, the wireless network must register the user in the new area and cancel its registration with the previous service provider if necessary. Typically there are different types of registrations and/or location update procedures. A service provider may divide its coverage area into different location areas. When a mobile user crosses the boundary of a location area, it is required to re-register with the network. The mobile phone can detect such a boundary change based on the location area identifier it receives from the system information broadcasted by a base station with which the mobile phone communicates.

2.1.3   Call Delivery

If a call is made to a roaming subscriber, the phone call is routed directly to the home MSC. The home MSC checks the HLR to determine the location of the subscriber. Roaming subscribers have their current visiting MSC ID stored in the HLR. So, the home MSC is able to route the incoming call to the visited network immediately. The home MSC is responsible for notifying the visiting MSC of an incoming call and delivering that call to the roamer. The home MSC first sends a route request to the visited MSC using the signaling network (most of the time this is a SS7 network). The visiting MSC returns a temporary directory number (TDN) to the home MSC, also via the signaling network. The TDN is a dynamically assigned temporary telephone number that the home MSC uses to forward a call via the PSTN. The incoming call is routed directly to the visiting MSC over the PSTN, through the home MSC.

2.1.4 Handoff

Sometimes, a mobile user may move out of the coverage area of the base station it is communicating with while he/she is on the phone. Thus, the cellular system needs to be able to handoff this user's phone to another base station before the wireless signals from the original base station degrade too much in order not to cause a disruption in the voice conversation. Normally, a cellular phone scans for the control signals from nearby base stations periodically. Such information is fed to the cellular network, e.g., MSC when the control signal from the base station it is communicating with drops below a certain threshold. The network will then initiate a handoff procedure with a nearby base station from which the phone can receive stronger radio signals so that that nearby base station can instruct the cellular phone to handoff the existing call to itself. This will typically require the cellular phone to tune to a new traffic channel. If the system is designed correctly, the mobile user should not be aware of such a handoff. However, when the system handoff parameters are not properly tuned, the mobile user either suffers a dropped call or can hear a short voice clip.

2.2 Traffic Routing in Wireless Networks

Traffic generated in a wireless network needs to be routed to its destination. The type of traffic carried by a network determines the protocol, and routing services that must be used. Two general routing services are provided by the networks: connection and connectionless services. In connection-oriented routing, the communication path between a source and destination is fixed for the entire message duration and a call set up procedure is required to dedicate network resources to both the calling and called parties. Connectionless services do not require a connection set up for the traffic but instead rely on packet-based transmissions. Packet switching is the most commonly used method to implement connectionless services. It allows many data users to remain virtually connected to the same physical channel in the network. A message can be broken into several packets, and each individual packet in a connectionless service is routed separately. Successive packets of the same message might travel completely different routes and encounter different delays throughout the network. Packets sent using connectionless routing do not necessarily arrive in the order of transmission and must be reordered at the receiver. Some packets may be lost due to network or link failure. However, redundancy can be built in to re-create the entire message at the receiver if sufficient packets arrive. Alternatively, only those lost packets need to be retransmitted rather than the whole message. However, this requires more overhead information for each

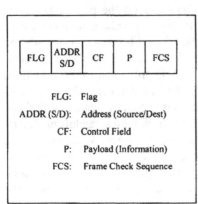

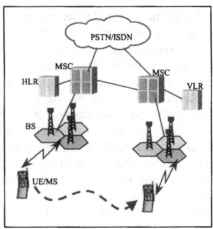

Figure 2-1 Typical Packet Format   *Figure 2-2* First-Generation Cellular Network

packet. Typical packet overhead information includes the packet source address, the destination address, and information to properly order packets at the receiver. Figure 2-1 shows an example of a transmitted packet with some typical fields, e.g., the flag bits, the address field, the control field, the information field, and the frame check sequence field. The flag bits are used to indicate the beginning and end of each packet. The address field contains the source and destination address of a message. The control field defines functions such as automatic repeat requests (ARQ), packet length, and packet sequencing information. The information field contains the user data of variable length. The final field is the frame check sequence (FCS) field or the Cyclic Redundancy Check (CRC) for error detection and/or correction.

2.3   First- and Second-Generation Cellular Radio Network

Figure 2-2 shows a diagram of a first-generation cellular radio network, which includes the mobile terminals, the base stations, and the Mobile Switching Centers.

First-generation wireless systems provide analog speech and inefficient, low-rate data transmission between the base station and the mobile user. The speech signals are usually digitized for transmission between the base station and the MSC. AMPS [You79] is an example of the first-generation wireless network which was first built by engineers from AT&T Bell Laboratories. In the first-generation cellular networks, the MSC maintains all mobile related information and controls each mobile handoff. The MSC also performs all of the network management functions, e.g., call handling and processing, billing, etc. The MSC is interconnected with the PSTN via wired trunks and a tandem switch. MSCs are also connected with other MSCs via dedicated signaling channels (mostly via SS7 network) for the exchange of location, authentication, and call signaling information.

The US cellular carriers use the IS-41 protocol [IS41] to allow MSCs of different service providers to pass information about their subscribers to other MSCs on demand. IS-41 relies on the autonomous registration feature of AMPS. A mobile uses autonomous registration to notify a serving MSC of its presence and location. The mobile accomplishes this by periodically transmitting its identity information, e.g., MIN and ESN, which allows the MSC to constantly update an entry in its database about the whereabouts of the mobile. The MSC is able to distinguish home users from roaming users based on the MIN of each active user. The Home Location Register (HLR) keeps the location information of each home subscriber while the Visiting Location Register (VLR) only keeps information of a roaming user. The visited system creates a VLR record for each new roamer and notifies the home system via the IS-41 so it can update its own HLR.

Second-generation wireless systems use digital modulation and provide advanced call processing capabilities. Examples of second-generation wireless systems include the TDMA and CDMA US digital standards (e.g., Telecommunication Industry Association IS-136 and IS95 standards), and the Global System for Mobile (GSM). In second-generation wireless systems, a base station controller (refer to Figure 2-3) is inserted between the base stations and the MSC to reduce the computational burden of the MSC. Dedicated control channels are provided within the air interface for exchanging voice and control information simultaneously between the subscriber, the base station, and the MSC while a call is in progress. Second-generation wireless networks were also designed to provide paging, facsimile, and higher-data rate network access. In addition, the mobile units perform more tasks to assist in handoff decision, e.g., reporting of received power and adjacent base station scanning. Second-generation mobile units can also perform data encoding and encryption.

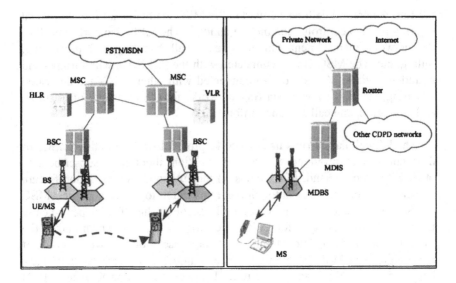

Figure 2-3. Second-Generation Cellular Network   *Figure 2-4.* CDPD Network Architecture

2.4   Deficiencies of First- and Second-Generation Wireless Systems

First-generation cellular systems provide connection-oriented services for each voice user. Voice channels are dedicated to the users at a serving base station and network resources are dedicated to the voice traffic on initiation of a call. The MSC sets up a dedicated voice channel connection between the base station and the PSTN for the duration of a cellular phone call. Circuit switching is used to transmit voice traffic to/from the user's terminal to the PSTN. Circuit switching establishes a dedicated radio channel between the base station and the mobile, and a dedicated phone line between the MSC and the PSTN for the entire duration of a call.

First-generation cellular systems provide data communications using circuit switching. Wireless data services such as fax and electronic mail are not well supported by circuit switching because of their short, bursty transmission, which are followed by periods of inactivity. Often, the time required to establish a circuit exceeds the duration of the data transmission. Modem signals carrying data need to be passed through the audio filters that are designed for analog, FM, and common air interfaces. Thus, it is both clumsy and inefficient, e.g., voice filtering must be deactivated when data are transmitted.

2.5 Second-Generation Cellular Networks Offering Wireless Data Services

In 1993, the US cellular industry developed the Cellular digital packet data (CDPD) standard to co-exist with conventional voice-only cellular system. CDPD is a data service for first- and second-generation US cellular systems and uses a full 30-KHz AMPS channel on a shared basis [Ken97]. Similarly, General Packet Radio Service (GPRS) was developed by the European standard body 3GPP to provide data service over GSM networks. Below, we provide a brief summary of CDPD.

CDPD provides mobile packet data connectivity to existing data networks and other cellular systems without any additional bandwidth requirements. CDPD directly overlays with the existing cellular infrastructure and uses existing base station equipment, making it simple and inexpensive to install. However, CDPD does not use the MSC for traffic routing. The active users are connected through the mobile data base stations (MDBS) to the Internet via intermediate systems (MD-IS) which act as servers and routers for the data users as shown in Figure 2-4.

CDPD often uses the same base station antenna as an AMPS cell and the same cellular voice RF plans. Normally, one or more 30KHz channels are dedicated to CDPD in each sector of the cell site. Alternatively, CDPD radios can share 30-KHz channels with AMPS calls, where AMPS calls have priority over data. This is referred to as channel hopping. The new logical entity included in the CDPD network architecture, called the Mobile Data Intermediate System, supports CDPD mobile protocols, e.g., authentication of mobile terminals, mobility management, accounting, and interservice provider interface, connections to wide area networks, etc.

Simulation results reported in [Ken97] indicate that the maximum CDPD throughput that a user can expect is about 19 Kbps. A similar limitation exists for GPRS. However, as wireless data users become more sophisticated, small airlink channel becomes insufficient to meet users' demand. Wireless data users are interested in more advanced applications, e.g., downloading music and video clips while they are on the move. They desire to have bigger airlink pipes and quality of service (QoS) features from the wireless networks.

2.6   Third-Generation Wireless Networks and Wireless LANs

The deficiencies of the first- and second-generation wireless systems prevent them from allowing roaming users to enjoy high data rate connections and multimedia communications. The aim of third-generation wireless networks is to introduce a single set of standards that provide higher airlink bandwidth and support multimedia applications. In addition, the third-generation wireless systems are expected to be able to communicate with other information networks, e.g., the Internet and other public and private databases. Examples of third-generation wireless systems are TIA 1xEV Data Only (or commonly referred to as High Data Rate system)-based networks [EVDO], TIA 1xEVDV-based networks [EVDV], and 3GPP UMTS networks [UMTS]. Such 3G systems promise a peak airlink bandwidth of 2–3Mbps. We will discuss CDMA2000 and UMTS networks in more detail in subsequent chapters.

Third-generation wireless networks are supposed to have enhanced voice capacity and are capable of supporting more sophisticated data applications with their rich Quality of Service features. However, since the airlink technology used for the European third-generation systems is completely different from the second-generation and the systems become more complex, the standardization effort has been delayed. Meanwhile, the IEEE 802.11-based wireless LAN system [80211] has become a mature technology, and hotspot services have begun to flourish in the United States. Currently wireless LAN systems can only cater to hotspots such as airports, big conference venues, and hotels. Limited roaming capabilities are provided. Thus, WLAN services can only complement wireless WAN services provided by third-generation systems. Wireless service providers are eager to see an integrated 3G and wireless LAN systems that can allow them the flexibility to direct hotspot traffic to wireless LAN systems whenever possible while supporting increasing number of voice subscribers using third-generation wireless systems.

Figure 2-5 [Per02] illustrates an integrated architecture of a 3G cellular and wireless LAN system. Typically in a wireless system, we have a radio access portion in which the mobile terminal communicates with a base station. The radio interface may terminate within the base station (referred to as an access point in a wireless LAN system) as in a wireless LAN system or terminate at a radio network controller as in a cellular system. In a wireless LAN system, the traffic will be routed from the base station to a centralized access controller via a wireless LAN gateway. Alternatively, the access

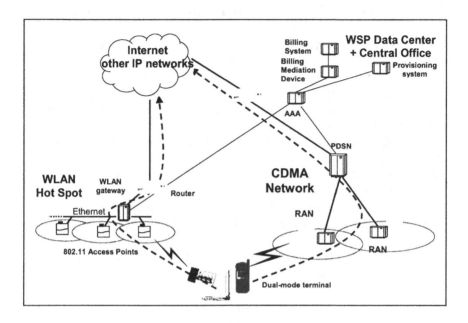

Figure 2-5. An Integrated 3G and Wireless LAN Network [Per02]

controller can also be co-located within the wireless LAN gateway. The access controller authenticates wireless LAN users and performs a simple admission control strategy. The first-generation wireless LAN system seldom performs sophisticated radio resource management since there are still not many wireless LAN users yet. For cellular system, a radio network controller performs radio resource management, user session management and mobility management. For cellular systems that utilize TDMA/GSM technology, the radio network controller is often referred to as the base station controller.

For CDMA/WCDMA technology, the radio network controller is the network element where the combining of the radio frames from a user terminal is performed to take advantage of the path diversity from different base stations to the user. From the radio network controller, the circuit and packet data traffic will be segregated and routed to different portions of the core network. The voice traffic will be carried to a mobile switching center while the data traffic will be carried via specialized routers that support mobility to the regular Internet.

2.7   Transport Choices for Wireless Backhaul Networks

The portion of the network that links the various base stations and radio network controllers together is often referred to as the radio access network while the portion of the network that comprises the gateway support nodes and mobile switching centers is referred to as the core network. The amount of traffic capacity required in either the radio access or core network is highly dependent on the type of traffic carried. For example, the voice traffic from a subscriber requires dedicated network access to provide real-time communications, whereas control and signaling traffic may be able to share network resources with other bursty users. Alternatively, some traffic may have an urgent delivery schedule while some may not need to be sent in real-time. The type of traffic carried by a network determines the routing services, protocols, and call handling techniques that must be employed.

Typically, the radio access networks can be point-to-point leased T1/T3 lines, frame relay networks, or ATM networks. The choice of the transport technology for the 3G radio access networks is often made based on whether or not the wireless service providers already have some existing cellular networks where the radio access networks are of a particular technology unless it is mandated by standard to use a particular technology. For third-generation networks such as UMTS, the standards (for UMTS only Release 99 standard) mandate that ATM be used as the transport technology for carrying voice and data traffic in the radio access networks. This is because the MAC protocol units between the base stations and radio network controllers need to meet tight delay/jitter requirements. ATM can negotiate the required QoS treatment and enforce it accordingly.

Most recently, some work has been done in MWIF, 3GPP2, and 3GPP to consider IP technology for radio access networks [MWIF], [25.933]. The primary advantages that such a transition offers to the service providers are the potential savings resulting from the less expensive IP equipment cost and the ease of maintenance due to the convergence of core and access networks. IP-based radio access networks allow several green-field wireless service providers to share a high bandwidth IP network by transporting their individual traffic through secured VPN tunnels. Such sharing of transport network allows the operators to reduce the recurring cost of leasing transport facilities. Since IP network does not provide stringent QoS capabilities, the task of transporting voice and other delay-sensitive traffic in an IP network remains an open challenging design task.

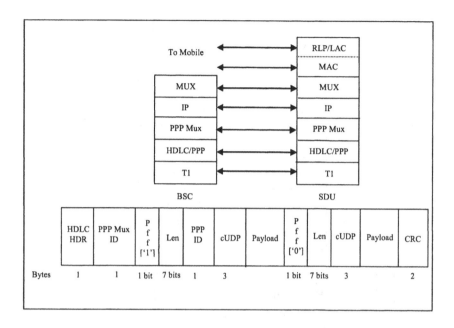

Figure 2-6 PPPMux - A Proposed Protocol Option for IP-Based RAN

The traffic that flows in the radio access network (RAN) is heterogeneous; some of the traffic types have stringent delay, loss, and jitter requirements. For example, the delay for transporting voice frames must be less than a few milliseconds and the jitter must be small for acceptable voice quality. For CDMA systems, performance enhancement schemes such as soft handoff also impose timely data delivery constraints on the RAN bearer traffic. On the forward link, traffic frames from different soft handoff legs need to reach the base station in time so that the mobile station can receive the frames synchronously. Similarly, on the reverse link, traffic frames need to arrive at the frame selector within predefined delays.

Several schemes have been proposed in various standards for IP-based radio access networks. Three of the proposed schemes are shown in Figures 2-6 to 2-8. The protocol stack for PPPMux [Paz01] is given in Figure 2-6. Several PPP encapsulated packets are sent within a single PPP frame. Thus, the PPP overhead per packet is reduced. Each PPP encapsulated frame is called a PPP subframe. A one-byte field, which consists of a PFF bit and a 7-bit length field is included. The PFF flag is set if PPP protocol ID is included within the PPP subframe. Only the first subframe needs to carry the PPP

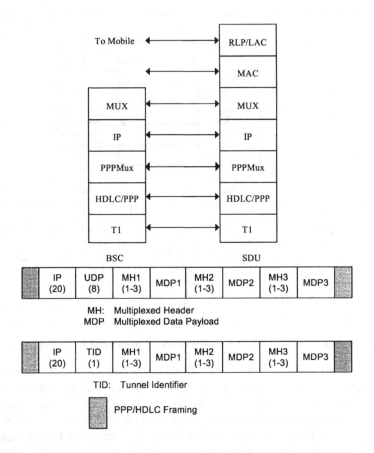

Figure 2-7 Another Proposed Protocol Option for IP-RAN: LIPE

Protocol ID. Each subframe contains a 3-byte compressed UDP/IP header and the data (either voice or data RLC Protocol Data Units). LIPE [Chu00] is a protocol proposed by Bell Laboratories researchers for multiplexing small voice/video/data frames into a single IP packet. LIPE uses either UDP/IP or IP as the transport layer. Its protocol stack and the encapsulated format are shown in Figure 2-7.

The MPLSMux scheme proposed in [Chu03a] is illustrated in Figure 2-8. Each subframe contains a header (LEN field) that conveys the length of the subframe. Several subframes can be multiplexed within an MPLS frame A typical MPLSMux multiplexing structure consists of a mandatory outer

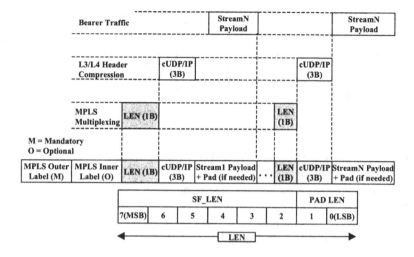

Figure 2-8 A Proposed Protocol Option for IPRAN: MPLSMux [Chu03a]©IEEE Copyright
2003

label, with zero or more inner labels, and one or more MPLSMux subframes
consisting of a one-octet header and variable length primary payload.

Simulation results comparing the voice carrying capacity of different
proposed transport options were presented in [Chu03a]. Table 2-1
summarizes the overhead values used in the simulation results and Table 2-2
provides the reported results. More results on the mixed voice and data
carrying capacity of the BTS-RNC links are discussed in [Sar03a], [Sar03b]
and a technique on how to improve such capacity is discussed in Chapter 7.

Table 2-1. Summary of the Overhead Values for Various Options

Protocol Type	Per Stream Overhead	Container Overhead
AAL2/ATM	3 bytes [AAL2]	6 bytes (5B ATM+1B STF)
LIPE	3 bytes [LIPE]	7 bytes (3B cUDP/IP+4B of PPP/HDLC)
PPPMux	4 bytes (3B cUDP+ 1B Len+Pff field)	5 bytes (1B HDLC+1B PPPMuxID + 1B PPPID+ 2B CRC)
MPLSMux	4 bytes (3B cUDP+ 1B Len field)	8 bytes (4B MPLS+ 4B PPP/HDLC)
MPLS	11 bytes (3B cUDP/IP+ 4B MPLS Label+4B PPP/HDLC) - same per stream and container OH	

Table 2-2. Comparison of Voice Carrying Capacity Using Different Multiplexing Options

Number of Voice Users Per T1				
AAL2/ATM	MPLS	PPPMUX	MPLSMux	LIPE
127	122	135	133	142

2.8   End-to-End Protocol Stack

In this section, we present examples of the end-to-end protocol stacks for both circuit switched and packet switched services.

2.8.1   Circuit Switched Service

Figure 2-9 presents an end-to-end protocol stack for circuit switched service in UMTS. The bearer traffic (voice and circuit data) is carried over transparent or unacknowledged mode RLC/MAC over physical radio connections to the base station. The base station carried RLC/MAC protocol data units over AAL2/ATM to RNC. The various radio frames are soft-combined within the frame selector in RNC and then the best voice frame is carried again via AAL2/ATM to MSC (denoted as core network (CN) in Figure 2-9). The call control (CC), mobility management (MM), and session management (SM) procedures are performed between a User Equipment (UE) and the Core Network. The RRC connection is established between a UE and an RNC. The signaling connection between a NodeB and an RNC for carrying signaling messages between the NodeB and the RNC is referred to as the NBAP connection. Similarly, the RANAP connection is established between an RNC and the core network (e.g., MSC) to carry signaling messages between the RNC and the core network.

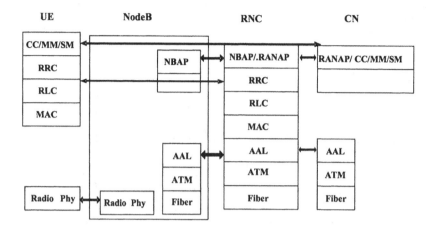

Figure 2-9 End-to-End Protocol Stack for Circuit Switched Service

2.8.2   Packet Data Service

Figure 2-10 shows an end-to-end protocol stack for packet data service for a UMTS system. In this example, we show how an end-to-end VPN tunnel can be created to carry a user's IP packets securely over the UMTS system. The IP packets generated by the user are encapsulated as PPP frames. The PPP frames are further encapsulated as PPTP [Ham99] protocol data units over the secured IP tunnels. The IP packets will be passed to a convergence layer (indicated as Packet Data Convergence Protocol (PDCP) in Figure 2-10). Note that if end-to-end VPN tunnels are not used, then the IP packets created by the application will be passed directly to the PDCP layer without going through the PPP/PPTP/IP layers. At the PDCP layer, the packets will be segmented into appropriate radio link protocol units. The radio link protocol units will then be passed to the medium access layer, where they are further turned into medium access protocol units. The MAC packets are then passed to the physical radio layer and transmitted over the air to the base station (denoted as NodeB in UMTS). From the base station, the MAC packets may be sent directly to the serving radio network controller or via another radio network controller (often referred to as a controlling RNC). The serving radio network controller is the network element that handles the radio resource management of that user session. The controlling RNC is the network element that controls the radio resources of a base station. Note that some encapsulating protocols are used to carry the MAC protocol units from

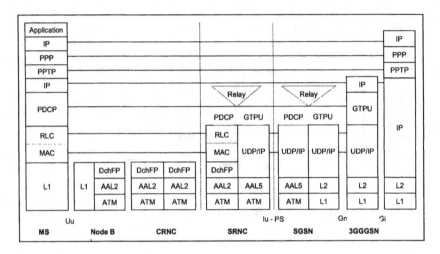

Figure 2-10. End-to-End Protocol Stack for PPTP User in 3G Networks

the base station to the serving radio access network. The serving radio network controller is the network element where the RLC and MAC terminate. The RLC protocol units are extracted from the MAC protocol units and reassembled. The extracted IP packets are passed to higher layers. The RLC protocol is used so that retransmissions can be performed to reduce the packet error rate seen at the IP layer between the user's terminal and RNC. Such retransmissions hide the error-prone airlink and provide reasonable end-to-end throughput to the user. From SRNC, the data packets are carried over special UDP tunnels to the Serving GPRS Support Node as well as the Gateway GPRS Support Node. The special UDP tunnels are referred to as GTP tunnels in UMTS. The GTP tunnels provide tunnel identifiers that allow the gateway support nodes to identify packets from different users' sessions. The GTP tunnels also allow the gateway support nodes to support different QoS classes for traffic from different users. Further details of the GTP tunnels are discussed in Chapter 8.

2.9   RLC/MAC Functions

Most of the data applications that mobile users run are built on top of TCP/IP[Ste94]. Reliable transport protocols such as TCP have been tuned for traditional networks with wired links and stationary hosts. These protocols assume congestion in the network is the primary cause for packet losses and unusual delays. TCP performs well over such networks by adapting to end-to-end delays and congestion losses. TCP reacts to packet losses by dropping its transmission window size before retransmitting packets, initiating congestion control or avoidance mechanisms (e.g., slow start [Jac88]), and backing off its retransmission timer (Karn's algorithm [Kar91]). Such measures result in a reduction in the load on the intermediate links, thereby controlling the congestion in the network. Unfortunately, when packets are lost for reasons other than congestion, such measures result in unnecessary reduction in end-to-end throughput. Communication over wireless links is often characterized by sporadic high bit-error rates, and intermittent connectivity due to handoffs. TCP performance in wireless networks suffers from significant throughput degradation and very large delays [Rac95].

There are three different approaches to improving TCP performance in such lossy systems: (1) the end-to-end approach, (2) the split-connection approach, and (3) the link-layer approach. In the end-to-end approach, selective acknowledgements are used to allow the sender to recover from multiple packet losses without resorting to a coarse timeout. An example of this approach is the SMART scheme [Kes96]. SMART uses acknowledgments that contain the cumulative acknowledgment and the sequence number of the packet that caused the receiver to generate the acknowledgment. The sender uses this information to create a bit-mask of packets that have been delivered successfully to the receiver. When the sender detects a gap in the bit-mask, it immediately assumes that the missing packets have been lost without considering the possibility that they simply may have been reordered. An example is shown in Figure 2-11. In Figure 2.11, the TCP stack in MT is enhanced to support selective acknowledgments. When packets 4 and 5 are received, the MT concludes that packets 2 and 3 are missing and starts sending selective acknowledgments indicating that packets 2 and 3 are missing. In this example, it takes the sender at least a one-way end-to-end delay to find out that the receiver does not receive certain packets.

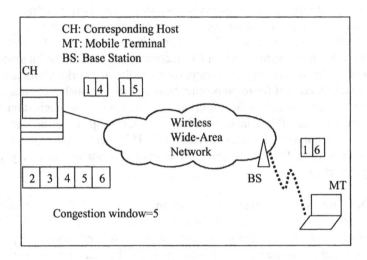

Figure 2-11 TCP with SMART-Based Selective Acknowledgments

The authors in [Har95b] combined the SMART scheme with the IETF selective acknowledgments [Mat96]. The sender retransmits a packet when it receives a SMART acknowledgment only if the same packet was not retransmitted within the last round-trip time. If no further SMART acknowledgments arrive, the sender falls back to the coarse time-out mechanism to recover from the loss. In addition, the author in [Har95b] also proposes an Explicit Loss Notification (ELN) option to TCP acknowledgments. When a packet is dropped on the wireless link, future cumulative acknowledgments corresponding to the lost packet are marked to identify that a non-congestion related loss has occurred. On receiving this information with duplicate acknowledgments, the sender may perform retransmissions without invoking the associated congestion-control procedures.

The second approach is the split-connection approach [Bak95] which completely hides the wireless link from the sender by terminating the TCP connection at the base station and uses a separate reliable connection between the base station and the destination hosts. The second connection can use techniques such as negative or selective acknowledgments to achieve better performance over the wireless links [Har95a]. An example of the split connection is shown in Figure 2-12. In Figure 2-12, we see that packets 2 to 6 are received at the base station but the mobile terminal has not received packet 2 so it keeps sending acknowledgements for packet 1. In this example, the base station does not implement SMART-based selective

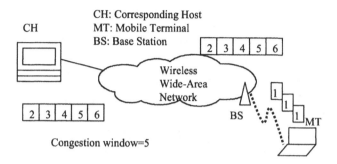

Figure 2-12 Split TCP Connection

acknowledgments so it does not know if MT has received other packets, e.g., packets 3, 4, 5, 6. One can always combine the split TCP approach with SMART-based selective acknowledgments at the airlink segment.

The third approach is the link-layer protocol which attempts to hide link-related losses from the TCP sender by using local retransmissions, e.g., automatic repeat request (ARQ) [Des93], [Aya95], and perhaps forward error correction [Lin83] over the wireless link. Airlink retransmissions help to achieve good TCP performance over wireless links [Har95b]. The link layer protocols for the digital cellular systems in the United States, both CDMA [Kha00] and TDMA [Nan98], primarily use ARQ techniques. While the TDMA radio link control (RLC) protocol guarantees reliable, in-order delivery of link layer frames, the CDMA radio link control protocol makes only a limited attempt and leaves eventual error recovery to the reliable transport layer. An example of the link-layer approach with SMART-based selective acknowledgement (ACK) is shown in Figure 2-13. In this example, we show that the base station and the MT use the SMART-based selective acknowledgment approach to do local retransmissions. Should the local retransmissions fail to recover the packet in time, then recovery is performed via the end-to-end acknowledgment. The base station can further suppress duplicate end-to-end ACKs.

To appreciate the usefulness of link layer retransmission, let us use a simple example to evaluate the probability that a TCP packet will be lost if no link-layer retransmission is performed compared to the case where link layer retransmission is performed. Assume that a TCP packet is divided into 25 link-layer packets at the base station before being transmitted over to the

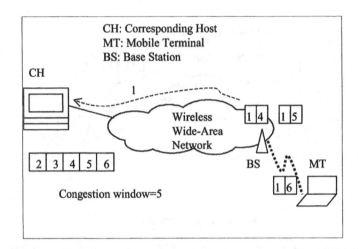

Figure 2-13 Link Layer with SMART-Based Selective Acknowledgments

mobile terminal. All of the 25 link layer packets must be received correctly before the original TCP packet can be delivered to the receiver. Otherwise, all the link-layer packets received will be dropped. If the probability of link-layer packet error is 10%, then, the probability that the TCP packet will be dropped is $0.928 = 1 - (0.9)^{25}$. If each link layer packet will be retransmitted at least once, then the probability of that TCP packet being dropped reduces to $0.222 = 1 - (0.99)^{25}$. Table 2-3 tabulates the packet drop probability as a function of the number of link-layer packet retransmissions assuming that the link-layer packet error rate is 10%.

Table 2-3. Packet Drop Probability with Different Number of Retransmissions

No of Retransmissions	Packet Drop Probability
0	0.928
1	0.222
2	0.0247
3	0.0025
4	0.00025
5	0.0000025
6	0.00000025

The extensive simulation studies reported in [Har95b] reveal that a reliable link-layer protocol that uses the knowledge of TCP to shield the sender from duplicate acknowledgments arising from wireless losses gives a 10–30% higher throughput than one that operates independent of TCP and does not attempt in-order delivery of packets. The split connection approach with standard TCP used for wireless hop shields the sender from wireless losses. However, the sender often stalls due to timeouts on the wireless connection, resulting in poor end-to-end throughput. A SMART-based selective acknowledgment mechanism for the wireless hop yields good throughput but the throughput is still slightly less than that for a well-tuned link-layer scheme that does not split the connection.

How RLC/MAC parameters affect end-to-end throughput for CDMA/UMTS networks is discussed in great detail in Chapter 9.

Acknowledgments

The wireless backhaul performance work reported in Section 2.8 was performed jointly with former Bell Laboratories colleagues Dr S. Abraham and K. Medeppalli.

2.10   Review Exercises

1. Give some examples of control information that is transferred to and from a cellular phone but is not necessary in conventional telephones. Why do cellular systems have separate control channels and traffic channels?

2. Why do cellular phone systems require the handoff procedure?

3. Name the additional network elements that are added to the AMP system to offer CDPD packet data services.

4. Assume that in a sector with a number of c AMPS channels, calls that originate in the sector and calls that are handed off to a sector are generated in accordance with a Poisson process with normalized rate of calls/unit. The length of time an AMPS call holds onto an AMPS channel is assumed to be uniformly distributed with unit mean. Use arguments from renewal theory to determine the number of idle channels per sector. In addition, calculate the average length of time an AMPS channel is idle by assuming that each channel receives an equal fraction of the AMPS call load.

5. Assuming voice packets follow the packet size distribution listed below, determine the voice call carrying capacity of different IP-based multiplexing schemes on a T1 link assuming that the link can only be loaded up to 90%.

Probability	Size (Bytes)
0.291	34
0.039	16
0.072	7
0.598	3

2.11   References

[25.933] 3GPP TSG RAN V.2.0.0, "IP Transport in UTRAN Work Task Technical Report," May 2001.

[802.11] IEEE Standard 802.11, "Part 11: Wireless LAN Medium Access Control & Physical Layer Specifications," June 1997.

[Aya95] E. Ayanoglu et al, "AIRMAIL: A link-layer protocol for wireless networks," ACM/Baltzer Wireless Networks Journal, 1:47–60, Febuary 1995.

[Bak95] A. Bakre and B. Badrinath, "I-TCP: Indirect TCP for Mobile Hosts," Proceedings of 15th International Conference on Distributed Computing Systems (CDCS), May 1995

[Chu00] M. Chuah, E-H. Valencia, "A Light Weight IP Encapsulation Scheme," IETF draft submitted to AVT Working Group, June 2000.

[Chu03a] M. Chuah, K. Medepalli, S. Park, "MPLS-Mux: An efficient protocol for providing QoS in IP-based Wireless Networks," Proceedings of VTC, 2003.

[Des93] A. Desimone, M. Chuah, O. Yue, "Throughput Performance of Transport-Layer Protocols over Wireless LANs," Proceedings of Globecom, 1993.

[EVDO] P. Bender et al, "A Bandwidth Efficient High Speed Wireless Data Service for Nomadic Users," IEEE Communications Magazine, July 2000.

[EVDV] L. Hsu, et al, "Evolution Towards Simultaneous High-Speed Packet Data and Voice Services: An overview of cdma2000 1xEV-DV," Proceedings of VTC, 2003.

[Ham99] K. Hamzeh, et al, "Point-to-Point Tunneling Protocol," RFC2637, July 1999.

[Har95a] H. Balakrishnan, S. Seshan, and R. Katz, "Improving reliable Transport and Handoff Performance in Cellular Wireless Networks," ACM Wireless Networks, 1(4), December 1995.

[Har95b] H. Balakrishnan, V. Padmanabhan, S. Seshan and R. Katz, "A comparison of mechanisms for improving TCP performance over wireless links," ACM/IEEE Transaction on Networking, 1995.

[IS41] TIA, "Cellular Radio-Telecommunications Intersystem Operations," Interim Standard 41B, December 1991.

[Jac88] V. Jacobson, "Congestion Avoidance and Control," Proceedings of ACM Sigcomm, 88, August 1988.

[Kar91] P. Karn et al, "Improving Round-Trip Time Estimates in Reliable Transport Protocols," ACM Transactions on Computer Systems, Vol 9(4): 364–373, November 1991.

[Ken97] K. Budka, H. Jiang, S. Sommars, "Cellular Digital Packet Data Network," Bell Laboratories System Technical Journal, Summer of 1997.

[Kes96] S. Keshav and S. Morgan, "Smart retransmission: Performance with overload and random losses," 1996.

[Kha00] F. Khan, S. Kumar, S. Nanda, "TCP performance over CDMA2000 RLP," Proceedings of Vehicular Technology Conference, 2000

[Lin83] S. Lin and D.J. Costello, *Error Control Coding: Fundamentals and Applications*, Prentice Hall, Inc, 1983.

[Mat96] M. Mathis et al, "TCP Selective Acknowledgments Options," RFC2018, 1996.

[MWIF] MTR-006, "IP in the RAN as a transport option in 3rd generation mobile systems," MWIF Technical Report, June 2001.

[Nan98] S. Nanda, R. Ejzak, B. Doshi, "A retransmission scheme for circuit-mode data on wireless links," IEEE Journal on Selected Communications, Vol 12, No 8, October 1994.

[Paz01] R. Pazhayannur et al, "PPP Multiplexing," RFC 3153, August 2001.

[Per02] D. Benenati et al, "A Seamless Mobile VPN Data Solution for CDMA2000, UMTS and WLAN Users," Bell Labs Technical Journal, December 2002.

[Rac95] R. Caceres et al, "Improving the Performance of Reliable Transport Protocols in Mobile Computing Environments," IEEE Journal on Selected Areas in Communications, 13(5), June 1995.

[Sar03a] C. Saraydar, S. Abraham, M. Chuah, A. Sampath, "Impact of Rate Control on the capacity of an Iub link: single service case," Proceedings of ICC, 2003.

[Sar03b] C. Saraydar, S. Abraham, M. Chuah, "Impact of Rate Control on the capacity of an Iub link: mixed service case," Proceedings of WCNC, 2003.

[Ste94] W. Stevens, *TCP/IP Illustrated*, Vol 1, Addison-Wesley, Reading, MA, November 1994.

[UMTS] K. Richardson, "UMTS Overview," Electronics and Computer Engineering Journal, June 2000, pp 75–101.

[You79] W. R. Young, "Advanced Mobile Phone Service: Introduction, Background and Objectives," Bell Systems Technical Journal, Vol 58, pp 1–14, January 1979.

Chapter 3

INTRODUCTION TO TRAFFIC ENGINEERING

3. INTRODUCTION

In second-generation cellular network, any user can enjoy at most on the average about 10–40 Kbps bandwidth. 3G networks are designed to provide a higher bit rate for any user. Higher bit rates facilitate some new services, e.g., video telephony and quick downloading of data. Compared to 2G networks such as GSM, 3G networks allow a user to negotiate the properties of a radio bearer. Attributes that define the characteristics of the radio bearer service may include throughput, transfer delay, and data packet error rate.

In Sections 3.1–3.3, we use UMTS as an example to show how Quality of Service (QoS) features are provided in 3G networks and how QoS engineering is used to support multimedia applications. UMTS allows a user to negotiate bearer characteristics that are most appropriate for an application. It also allows the bearer properties to be renegotiated during an active session. Bearer negotiation is initiated by a user while bearer renegotiation can be triggered either by a user or by the network.

The layered architecture of a UMTS bearer service [23.107] is depicted in Figure 3-1. Each bearer service on a specific layer offers its individual services using those provided by the layers below. The end-to-end service comprises three components: the Terminal Equipment (TE)/Mobile Terminal (MT) local bearer service, the UMTS bearer service, as well as the external bearer service. The UMTS bearer service consists of two parts: the Radio Access bearer service that provides transport of signaling or user data between a MT and a Serving GPRS Support Node (SGSN), and the Core Network (CN) bearer service that provides transport of signaling or user data between a SGSN and a Gateway GPRS Support Node (GGSN). The radio access bearer service is further realized by the radio bearer service (between a MT and an RNC) and the Iu bearer service (between an RNC and a SGSN). The radio bearer service makes use of the UTRA FDD/TDD

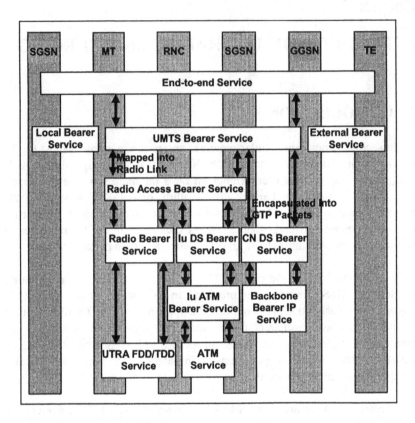

Figure 3-1 UMTS Bearer Service Architecture

physical layer services. The Iu Bearer Service is mapped into the physical transport bearer service within the radio access network, e.g., an ATM bearer service (for an ATM-based radio access network) or IP over ATM bearer service (for an IP-based radio access network). The IETF Differentiated Service (DS) feature can be used to provide QoS at the Core Network (CN) Bearer Service.

3.1   QoS Requirements of Internet Applications

There are several popular Internet applications, e.g., WWW-browsing, emails, streaming video and audio, etc. These applications have different QoS requirements. Some are tolerant to packet errors and packet losses while others are not. In addition, some applications are tolerant to delay while others are not. Table 3-1 shows qualitatively how different UMTS

QoS classes can be used to meet the QoS requirements of these different applications. Conversational voice and video can tolerate some packet errors but have stringent delay requirements. Applications such as online games and telnet sessions cannot tolerate packet errors and also have small delay requirements. E-commerce and WWW-browsing applications can tolerate slighter higher delay but are also not tolerant to packet errors. Email and FTP are applications that can tolerate even more delay than WWW-browsing.

3.2   UMTS QoS Classes

In UMTS, four traffic classes have been defined:
1. Conversational
2. Streaming
3. Interactive
4. Background classes

Each UMTS QoS class can be described using the QoS attributes that have certain ranges. The defined UMTS QoS attributes can be classified into three groups:

1. Delay attributes: transfer delay
2. Bandwidth attributes: maximum bit rate, guaranteed bit rate
3. Reliability attributes: delivery order, traffic handling priority, allocation/retention priority, etc.

The main distinguishing factor among these classes is the delay sensitivity required by each traffic class. Conversational class is very delay sensitive while background class is delay insensitive. Examples of the attributes and value ranges for different UMTS bearer services [23.107] are shown in Tables 3-2 and 3-3. The values for residual Bit Error Rate (BER) and SDU error ratio are derived from different CRC lengths on layer 1.

Table 3-1. Qualitative QoS Requirements for Different Applications

Error tolerant	Conversational voice and video	Voice messaging	Streaming audio and video	Fax
Error intolerant	Telnet, interactive games	E-commerce, WWW browsing,	FTP, still image, paging	Email arrival notification
	Conversational (delay << 1 sec)	Interactive (delay approx. 1 sec)	Streaming (delay <10 sec)	Background (delay >10 sec)

Table 3-2. UMTS QoS Requirements

Medium	Application	Degree of symmetry	Data rate	Key performance parameters and target values		
				End-to-end one way delay	Delay variation within a call	Informatio n loss
Audio	Conversational voice	Two-way	4-25 Kb/s	<150 ms preferred <400 ms limit	<1 ms	<3%FER
Video	Video phone	Two-way	32-384 kb/s	<150 ms preferred <400 ms limit. Lip-synch. <100ms		<1%FER
Data	Telemetry-two-way control	Two-way	<28.8 Kb/s	<250 ms	N/A	Zero
Data	Interactive games	Two-way	< 1Kbps	<250 ms	N/A	Zero
Data	Telnet	Two-way (asymmetric)	<1 Kbps	<250 ms	N/A	Zero

Table 3-3. UMTS Bearer Service Attributes and Value Range [23.107]

Traffic class	Conversational class	Streaming class	Interactive class	Background class
Maximum bit rate (Kbps)	< 2048	< 2048	< 2048	< 2048
Delivery order	Yes/No	Yes/No	Yes/No	Yes/No
Maximum SDU size (octets)	≤ 1500 or 1502	≤ 1500 or 1502	≤ 1500 or 1502	≤1500 or 1502
Delivery of erroneous SDUs	Yes/No	Yes/No	Yes/No	Yes/No
Residual BER	$5 \cdot 10^{-2}$, 10^{-2}, $5 \cdot 10^{-3}$, 10^{-3}, 10^{-5}, 10^{-6}	$5 \cdot 10^{-2}$, 10^{-2}, $5 \cdot 10^{-3}$, 10^{-3}, 10^{-4}, 10^{-5}, 10^{-6}	$4 \cdot 10^{-3}$, 10^{-5}, $6 \cdot 10^{-8}$	$4 \cdot 10^{-3}$, 10^{-5}, $6 \cdot 10^{-8}$
SDU error ratio	10^{-2}, $7 \cdot 10^{-3}$, 10^{-3}, 10^{-4}, 10^{-5}	10^{-1}, 10^{-2}, $7 \cdot 10^{-3}$, 10^{-3}, 10^{-4}, 10^{-5}	10^{-3}, 10^{-4}, 10^{-6}	10^{-3}, 10^{-4}, 10^{-6}
Transfer delay (ms)	100 – maximum value	250 – maximum value		
Guaranteed bit rate (Kbps)	< 2048	< 2048		
Traffic handling priority			1, 2, 3	
Allocation/Retention priority	1, 2, 3	1, 2, 3	1, 2, 3	1, 2, 3

3.2.1   Conversational Class

Conversational class is defined for real-time conversation that requires low end-to-end delay. The best-known example is speech service over circuit-switched bearers. With the popularity of the Internet and emerging needs for multimedia communications, a number of new applications also require this class of service, e.g., Voice over IP and video telephony. Subjective evaluations have shown that the end-to-end delay has to be less than 400 ms for video and audio conversation.

In UMTS, the speech codec used will be the Adaptive Multi-Rate (AMR) codec. Eight source rates are defined for this multi-rate speech codec: 12.2, 10.2, 7.95, 7.40, 6.70, 5.9, 5.15, and 4.75 Kbps. The AMR rates are controlled by the radio access network. The 12.2 Kbps AMR speech codec is the GSM EFR codec, the 7.4 Kbps codec is the US-TDMA speech codec, and the 6.7-Kbps codec is the Japanese PDC codec. The AMR specification provides error concealment. Thus, the AMR speech codec can tolerate about 1% Frame Error Rate (FER) of class A bits without any deterioration of speech quality. For Class B and C bits, a higher FER is allowed.

Video telephony is another application that will use conversational class for UMTS transport. The Bit Error Rate requirement for video telephony is more stringent than that of speech. In UMTS, H.324M has been chosen to be used for video telephony in circuit-switched connections [26.110].

3.2.2   Streaming Class

Multimedia streaming allows data to be transferred such that they can be processed as a steady and continuous stream. With streaming, the client browser can start displaying the data before the entire file has been transmitted. Streaming applications are very asymmetric and therefore typically can withstand more delay and jitter than conversational services.

3.2.3   Interactive Class

Interactive class can be used for applications where an end-user (either a machine or a human) is online requesting data from remote equipment. Examples of such applications are web browsing, server access and database retrieval. In addition, emerging m-commerce applications, e.g., wireless auction and online games, also require interactive class service.

3.2.4   Background Class

Data applications such as emails, Short Message Service (SMS), and downloading of databases can be delivered in the background or during off-peak hours since such applications do not require immediate action. The delay may be seconds, tens of seconds, or even minutes. Traffic from such applications can be transported using UMTS background class service.

3.3   QoS Engineering

To support multimedia applications on a common packet-based network, QoS engineering is required. All service providers are interested in maximizing their revenue via offerings of high-value services while maintaining high utilization of their installed infrastructure. QoS engineering is an end-to-end issue. It must take into account the user terminal, the air interface, the radio access network, the core network, and external networks. QoS engineering involves all layers, namely physical, MAC, radio link, ATM-based or IP-based backhaul network, backbone network, and application layers.

The following are common techniques used for QoS engineering:
- Traffic characterization

Before any QoS engineering work can be performed, the traffic engineer needs to understand the different applications that will be used by the subscribers, the traffic arrival patterns of such applications, and the packet sizes/volume generated/received by such applications during a typical active session.

- Admission control and network resource allocation

In the network, there are limited resources, e.g., available power, airlink capacity, buffers, CPU processing power within network elements. Therefore, admission control modules exist within different network elements to grant permission for different users to use different segments of the network resources. Normally, users are admitted based on their average QoS requirements. Once admitted, network resource management algorithms are used to allocate resources to the admitted users, attempting to smooth out any transient congestion.

- Server/link scheduling and traffic policing/shaping

Service providers would like to offer different grades of services to the users so that traffic from users that pay more can get better treatment, e.g., they can enjoy smaller transfer delay or others' traffic will be discarded to make room for their traffic. Thus, server/link scheduling algorithms exist in different network elements within the 3G network. In addition, the network elements may negotiate for different QoS requirements with their transport networks, e.g., average and peak bandwidth and maximum traffic burst size. Thus, traffic from different network elements may need to be policed/shaped before it is fed to another downstream network element.

- QoS mapping between networks and different layers

Airlink packets are transported to upstream network elements via the transport networks, which may be Frame-relay based, ATM-based, or IP-based. To meet end-to-end QoS requirements, there must be proper QoS mapping between one layer and the next and between networks. As an example, the conversational traffic needs to be transported using Constant Bit Rate (CBR) traffic within the ATM-based radio access network. Such traffic may be transported using Internet Engineering Task Force (IETF) Expedited Forwarding (EF)-style QoS class [Jac99] across the IP-based core network before it arrives at its destinations.

- Service level agreements (SLA) and policy management

Users may negotiate for SLA with the carriers. Thus, the network elements need to have the capability to monitor the users' perceived service level to ensure that their negotiated SLA can be met. Many value-added services, e.g., corporate access, broadcast/multicast services from third parties, etc. will be provided by the carriers to attract more subscribers. Thus, policies exist in different network elements to authorize the users as well as the network elements in the usage of network resources. For example, each enterprise may have a list of forbidden web sites that their employees are not supposed to access via their corporate networks. Thus, access control polices need to be installed and managed at the edge routers or VPN gateways within the Point of Presence of the extended 3G networks.

- Radio channel selection and bandwidth allocation

Different 3G channels are available to cater to different needs. For example, dedicated channels (DCH) and common packet channels such as Fast access channels (FACH) and downlink shared channels are provided in UMTS. Many users can share an FACH channel while a dedicated channel is tied up by a user once it is allocated. Since there is a limit on the number of available dedicated channels, a service provider may arrange to have the users share a common packet channel when their traffic is small and switch them to a dedicated channel when their traffic builds up. Intelligent radio channel selection and bandwidth allocation strategies allow the service providers to cater to more users' data needs using the same resources.

● RF power control and rate control among users

Typically, CDMA technology is used in 3G network because of its high spectral efficiency. Users experience different fading conditions from time to time. Power control algorithms are used to ensure that the users can enjoy consistent airlink performance (in terms of packet error rates). Sometimes, the users' allocated rates are adjusted based on the fading conditions.

● RF resource setup and teardown strategies

Airlink resources are limited. Thus, it is normal practice to release precious airlink resources during the idle times of packet data sessions where users pause to digest the information they obtain before their next action. RF resource teardown procedures are performed to release airlink resources. When new packets of the same sessions arrive, RF resource set up procedures need to be performed so that such packets can be delivered without too much delay. Such a resource management scheme enables more users to share the airlink bandwidth, but it creates additional signaling messages. Thus, good RF resource setup/teardown strategies need to be designed to optimize the overall network performance.

● Network design based on traffic engineering rules

Designing a new 3G network or enhancing an existing 2/2.5G network to a 3G network is a complex process. The service providers need to have a reasonably good forecast of the needs of their current and prospective customers. Then, the service providers can estimate the potential traffic that will be generated by such subscribers and design their new network or evolve their current network to meet emerging demand. Traffic engineering rules will be used during the network design process to ensure that an appropriate number of network elements are chosen and placed intelligently in suitable geographical locations to meet subscribers' demands. For example, the base stations and base station controllers should be located appropriately such that the voice and data traffic generated by the subscribers can be carried within the radio access network using minimum network resources.

3.4 Traffic Modeling

Performance analysis tools can be built using a general and consistent traffic model. The generic traffic model framework should be able to evolve easily with the emergence of new data applications and services. A good understanding of the traffic patterns generated by potential subscribers allows the service providers to design a 3G network efficiently. Realistic traffic models are required to size various network elements within a 3G network. For example, if we assume that only the 12.2–Kbps rate will be used by the AMR codec and that no silent detection is used in a 3G voice network, then the total traffic generated by a voice user will be 13.3 Kbps (taking into consideration the ATM transport overhead). Similarly, a good traffic model also enables us to estimate the end-to-end user perceived throughput.

Using a WWW-browsing session as a specific example, we illustrate why good traffic modeling is important to help us understand the difference between the user-perceived throughput and the allocated airlink bandwidth. It also provides clues for us to design a more efficient radio resource management scheme.

Figures 3-2 and 3-3 show the traces of TCP connections for a simple (Net-Adobe-1) and a sophisticated (Net-CNN-1) Web page download over a 3G1x network. The *x*-axis plots the increasing session time (in seconds) while the *y*-axis plots the TCP port numbers observed during the session. The simple Web page download results in 12 TCP connections and has a size of 190 Kbytes. The average downlink and uplink bandwidth used is 62 Kbps and 8 Kbps respectively for the simple Web page. The peak downlink bandwidth observed within a second period is about 270 Kbps. The average downlink latency is about 120 ms while the average uplink latency is about 160 ms. The simple Web page download time is 24.5 seconds. The sophisticated Web page download results in 30 TCP connections and has a page size of 205 Kbytes.

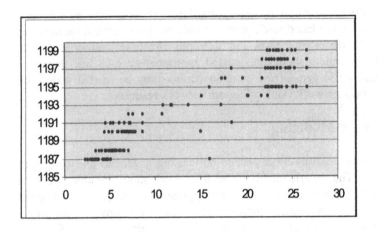

Net-Adobe-1
- Average downlink bandwidth: 62 Kbps
- Average uplink bandwidth: 8 Kbps
- Page download time: 24.5 s
- Average downlink packet latency: 120 ms
- Average uplink packet latency: 160 ms
- 12 TCP connections
- Total download bytes = 190,000 bytes
- Peak one-second downlink bandwidth: 270 Kbps

Figure 3-2 Simple Web Page Download[1]

The average downlink and uplink bandwidth is 125 Kbps and 17 Kbps, respectively. The peak downlink bandwidth observed within a 1-s period is about 250 Kbps. The average downlink latency is about 73 ms while the average uplink latency is 145 ms. The Web page download time is 13 s.

Without a good understanding of the interactions between the Web page size distribution, TCP, and the 3G1x network, one cannot explain the above observable performance. In a 3G1x network, a data user is given a 9.6 Kbps fundamental channel and an additional supplemental channel of varying bandwidth will be allocated based on the data burst size that arrives at the buffer within the base station controller. The peak airlink bandwidth that can be allocated is 153.6 Kbps. When TCP compression is used, one can typically see a compression ratio of 1.6–1.75:1. So, it is reasonable to see an observable peak bandwidth in the range of 250–270 Kbps.

---

[1] The plot in Figure 3-2 was provided by Lucent Technologies colleague Joe Courtier.

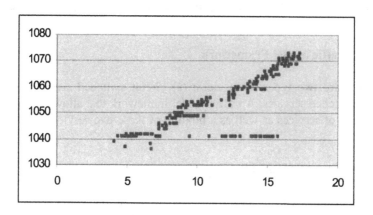

Net-CNN-1

- Average downlink bandwidth: 125 Kbps
- Average uplink bandwidth: 17 Kbps
- Page download time: 13 s
- Average downlink packet latency: 73 ms
- Average uplink packet latency: 145 ms
- 30 TCP connections
- Total download bytes = 205,000 bytes
- Peak one-second downlink bandwidth: 250 Kbps

Figure 3-3 Sophisticated Web Page Download[2]

A typical Web page has several hidden objects that spawn new TCP connections. There are a few large objects and many small objects on the Net-CNN-1 Web page. When the main object size is large, the window size of the relevant TCP connection has an opportunity to grow and hence results in better TCP throughput. Thus, even though there are 30 TCP connections, the average downlink bandwidth observed is larger for the Net-CNN-1 Web page.

The above example illustrates the importance of understanding traffic patterns generated by different Web pages so that one can explain the observed end-to-end throughput. By understanding the interactions between TCP applications and the lower layers, e.g., RLC/MAC, and physical layer,

---

[2] The plot in Figure 3-3 was provided by Lucent Technologies colleague Joe Courtier.

one can devise mechanisms to improve the user perceived performance over the wireless systems.

3.4.1   Traffic Model Framework

There are two types of services: (a) circuit switched services and (b) packet switched services. A traffic model framework that allows us to model both types of services as well as a mixture of these two services needs to be developed.

3.4.1.1 Circuit Switched Services

Two major circuit switched services are voice and video services. For voice calls, we need to make an assumption on the voice busy hour call attempt (BHCA) per subscriber and the average call holding time (ACHT). We also need to know the percentage distribution of different types of calls, e.g., mobile to PSTN calls, PSTN to mobile calls, and mobile-to-mobile calls. We need to understand whether the multi-rate AMR codec will be used or only the 12.2–Kbps AMR codec will be used. Similar statistics are required for video calls. For example, one can assume that the average bit rate that a video call used is 48 Kbps. In addition, we need to know the percentage of Public Switched Telephone Network (PSTN) calls that are answered, the percentage of call originations, pre-paid calls, etc.

3.4.1.2 Packet Switched Sessions

For packet switched sessions, we have a notion similar to busy call hour attempt that we term busy hour data session attempt (BHDSA). It gives us the number of session attempts per user during the busy hour. We also need to have the session duration statistics and session arrival distribution. Within a data session, we need to have statistics on the average number of transactions and its distribution, the mean user thinking time (idle time between transactions) and its distribution, and the average "ON" time (which includes the transaction time as well as inter-transaction time) and its distribution. In addition, we need to know the number of downlink/uplink packets within a data session, the packet size distributions, and the distribution for the packet inter-arrival times. Figure 3-4 shows the generic traffic model framework that we have described. One can determine the applicable traffic parameters for each application using this generic traffic model framework.

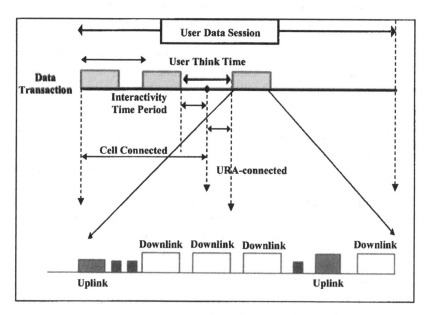

Figure 3-4 Generic Traffic Model

Some examples of the traffic parameter values of some popular Internet applications using the generic traffic model framework shown in Figure 3-4 are given below:

a) WWW-browsing

(i) Session interarrival time can be assumed to be exponentially distributed.

(ii) Number of transactions can be assumed to be normally distributed with a mean of 20.

(iii) User think time can be assumed to be gamma distributed with a mean of 40 s.

(iv) Number of downlink packets can be geometrically distributed with a mean of 40.

(v) Downlink packet size can be assumed to have discrete distribution with a mean of 820 bytes.

(vi) Number of uplink packets can be geometrically distributed with a mean of 40.

(vii) Uplink packet size can be assumed to have discrete distribution with a mean of 88 bytes.

(viii) One-way delay (between mobile and RNC) can be assumed to be exponentially distributed with a mean of 100 ms.

From these values, one can compute the average ON period, OFF period, and the session duration assuming that a 64–Kbps airlink bandwidth is allocated to the WWW-browsing user.

b) Email

(i) Session inter-arrival time can be assumed to be exponentially distributed

(ii) Each email session is assumed to consist of 10 email downloads, 7 reads, and 8 writes.

(iii) User think time is assumed to follow Pareto distribution with an average of 200 s for email writing or 80 s for email reading

(iv) The number of downlink packets is assumed to be geometrically distributed with a mean of 30

(v) Downlink packet size is assumed to have a discrete distribution with a mean of 1400 bytes.

(vi) The number of uplink packets is assumed to be geometrically distributed with a mean of 100.

(vii) Uplink packet size is assumed to follow a discrete distribution with a mean of 400 bytes.

(viii) One-way delay between the mobile and RNC is assumed to be exponentially distributed with a mean of 100 ms.

Again, assuming that a 64–Kbps airlink bandwidth is allocated, one can compute the average ON period, OFF period, and the session duration time.

3.4.2   Methodology for Traffic Characterization

In an operating network, traffic traces can be collected periodically for offline analysis. However, in a new network, fully integrated network elements may not be available for such traffic collection. Under such circumstances, normally an emulated system that mimics the real network will be built so that traffic traces can be collected. Here, we describe a methodology for traffic characterization using an emulated network with WAP application as an example.

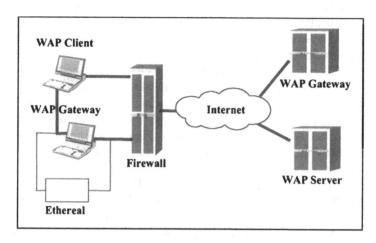

Figure 3-5 An Emulated WAP System

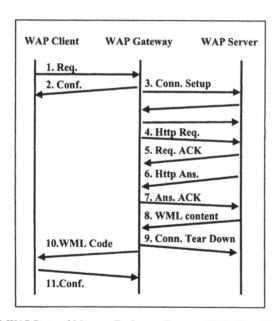

Figure 3-6. WAP Protocol Message Exchanges Between WAP Client/Gateway/Server

The WAP standard specifies a layered end-to-end communication protocol that allows users to receive news updates, stock updates, etc. A typical WAP system (as shown in Figure 3-5) consists of a WAP client that runs in a mobile mobile phone, a WAP gateway within a wireless network and a WAP Server in the Internet. The WAP client is connected to the WAP gateway via the wireless network. The bandwidth between the WAP client and the WAP gateway is artificially constrained to be the same airlink bandwidth in a typical 3G network, e.g., 9.6 Kbps, 19.2 Kbps, etc in 3G1x networks or 64 Kbps in UMTS networks. Figure 3-6 shows the protocol exchange between a WAP client, WAP gateway and a WAP server.

The WAP applications we used were: (a) BBC News trial service that gives headline, sports, show-biz, science technology news, etc; (b) weather updates; (c) sports news where scores and schedules of all kinds of matches, soccer, cricket, tennis can be found. Traces were collected from the emulated system. The collected traces were then analyzed to determine the packet size distribution for both downlink/uplink data packets, the user think time distribution, etc.

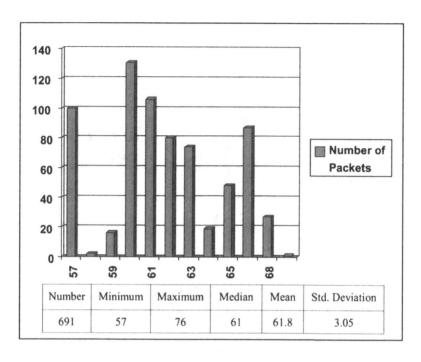

Figure 3-7 Uplink Packet Size Distribution for BBC News

In Figures 3-7 and 3-8, the *x*-axis is the packet size while the *y*-axis is the packet counts. The histogram for uplink packet size shown in Figure 3-7 indicates that the uplink packet size ranges from 57 to 76 bytes. Similarly, the histogram for downlink packet size shown in Figure 3-8 indicates that most of the downlink packet size ranges from 600 to 1100 bytes. Figure 3-9 shows the histogram for the user think time.

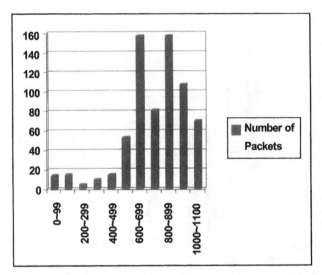

Number	Minimum	Maximum	Median	Mean	Std Deviation
691	57	76	61	61.8	3.05

Figure 3-8 Downlink Packet Size Distribution for BBC News

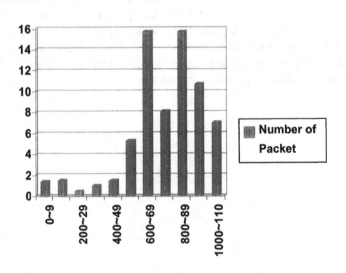

Figure 3-9 User Think Time Distribution

Once the traffic traces are collected, histograms are built. In addition, we fit an empirical distribution to each of the metrics we are interested in, e.g., packet size distribution. To fit an empirical distribution, we select a few reference distributions. We initialize the N distribution parameters by looking into the first N moments of the sampled data. Then, we tune the parameters. We check the fitness of the chosen empirical distribution using Quantile-to-Quantile (Q–Q) Plot and Complimentary Distribution Function (CDF) comparison.

For example, given the empirical distribution, $F(x)$ and a chosen reference distribution $G(x)$, we generate the Q–Q plot. The Q–Q plot is generated by choosing a few q_i such that $0 < q_i < 1$ and plotting $\{(x_i, y_i)$ such that $F(x_i) = G(y_i) = q_i\}$ (refer to Figure 3-10). If there is a good match between the two distributions, then the Q–Q plot is almost a straight line. Using this method and the information we collected for the downlink packet size (shown in Figure 3-11), we determined that a normal distribution with an average of 775 bytes and standard deviation of 180 bytes fits well with the empirical data.

Acknowledgments

The WAP traffic characterization efforts were carried out jointly with former Bell Laboratories colleague Dr. S. Strickland and a summer intern. The authors would like to thank 3GPP for giving permission to reproduce a table from the 3GPP TSG standard document [23.107]. 3GPP TSs and TRs are the property of ARIB, ATIS, ETSI, CCSA, TTA, and TTC who jointly own the copyright for them. They are subject to further modifications and are therefore provided to you "as is" for information purposes only. Further use is strictly prohibited.

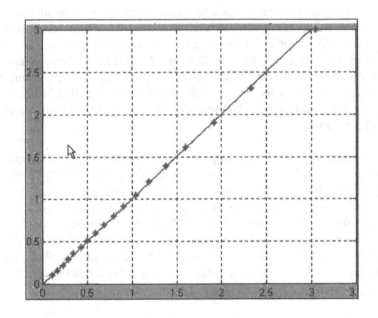

Figure 3-10. Q–Q Plot for Exponential Random Number Generator

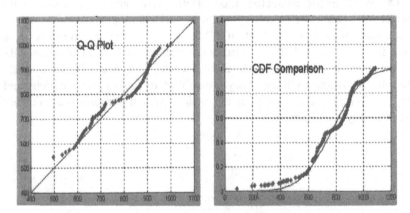

Figure 3-11. CDF Comparison for Downlink Packet Size Distribution

3.5 Review Exercises

1. Assume that the average downlink packet inter-arrival time is 50 ms. Compute the average ON period (which is defined as the average time it takes for all downlink packets within a transaction to arrive) using the parameters given for the Web application in Section 3.5.1.2. Then compute the session duration.

2. Repeat the above exercise using the parameters given for the email application.

3. Install ethereal on your PC or laptop and collect packet traces while you are surfing the web. Post-process the packet traces you have collected to extract information on packet size and packet inter-arrival times. Then use the Q-Q plot technique to fit a distribution to the packet size and the packet inter-arrival time.

3.6 References

[23.107] 3GPP, TSG Services and System Aspects, QoS Concept, 3G TS23.107, Version 1.3.0, 1999.

[26.110] 3GPP TSG Services and System Aspects, Codec for circuit switched multimedia telephony service: general descriptions, 3G TS 26.110 version 3.0.1, 1999.

[Jac99] V. Jacobson, et al, "An Expedited Forwarding PHB," RFC 2598, June 1999.

Chapter 4

OVERVIEW OF CDMA2000/UMTS ARCHITECTURE

4. INTRODUCTION

3G, the third-generation mobile communication standards drafted by the International Telecommunication Union (ITU), defines the next generation of mobile services that provide improved voice capacity, high-speed Internet, and multimedia services. ITU, working with industry bodies around the world, defines and approves technical and service requirements and standards under the IMT-2000 (International Telecommunication Union-2000) program. IMT-2000 is intended to introduce high-speed, broadband, and high-quality mobile multimedia telecommunication services "anytime, anywhere" to a worldwide market. The ITU requires that IMT-2000 or 3G networks deliver improved system capacity and spectrum efficiency over the 2G systems and support data services at minimum transmission rates of 144 Kbps in outdoor environment and 2 Mbps in indoor environments.

Based on these requirements, in 1999 ITU approved five radio interface technologies (as shown in Figure 4-1) for IMT-2000 standards as part of the ITU-R M. 1457 Recommendation. UMTS and CDMA2000 are two of the five standards.

IMT-2000 Terrestrial Radio Interface

IMT-2000 CDMA Direct Spread	IMT-2000 CDMA Multi-Carrier	IMT-2000 CDMA TDD	IMT-2000 TDMA Single Carrier	IMT-2000 FDMA/TDMA
WCDMA (UMTS)	CDMA2000 1X and 1xEV	UTRA TDD and TD-SCDMA	UWC-136/ EDGE	DECT

Figure 4-1. IMT-2000 Terrestrial Radio Interfaces Standards

4.1   Evolution of CDMA2000 Standards

CDMA2000 is an ITU approved, IMT-2000 or 3G standard, also known by its ITU name IMT-CDMA Multi-Carrier. The CDMA2000 specification was developed by the Third Generation Partnership Project 2 (3GPP2), which consists of five telecommunication standards bodies in North America and Asia: ARIB– Association of Radio Industries and Businesses in Japan, CCSA–China Communications Standards Association, TIA– Telecommunications Industry Association in North America, TTA– Telecommunications Technology Association in Korea, TTC– Telecommunications Technology Committee in Japan.

CDMA2000 is an evolution of an existing wireless standards based on IS-95. It supports 3G services as defined by the IMT-2000 and is fully backward compatible with IS-95. CDMA2000 is both an air interface and a core network solution for wireless operators to take advantage of the new market opportunities for mobile Internet. CDMA2000 represents a family of technologies that includes CDMA2000 1x and CDMA2000 1xEV.

- CDMA2000 1x:

    1x refers to the CDMA2000 implementation within the existing spectrum allocation for CDMAOne (trademark for IS-95), which uses a 1.25–MHz carrier. The technical term 1x means one times 1.25 MHz. CDMA2000 1x system can double the voice capacity of the CDMAOne networks and delivers a peak packet data rate of 307 Kbps in mobile environments. The enhancements include improved voice codec, better channel coding and modulation schemes, and enhanced power control techniques.

    The world's first 3G (CDMA2000 1x) commercial system was launched by SK Telecom of Korea, in October 2000. Since then, CDMA2000 1x has been deployed in Asia, North and South America, and Europe, and the subscriber base is growing everyday.

- CDMA2000 1xEV:

    1xEV stands for "1x Evolution." It includes 1xEV-DO (Data Only) and 1xEV-DV (Data and Voice). 1xEV-DO has already been standardized [Rev0][RevA] and the first commercial network was launched in 2002 by SK Telecom and KT Freetel in Korea. The 1xEV-DO system can deliver a peak data rate of 2.4 Mbps in the forward direction and 153.6 Kbps in the reverse direction. The enhancements include dynamic rate control, adaptive coding and modulation, hybrid ARQ schemes, and fast scheduling to optimize data transmission and delivery. It also supports multimedia applications such as MP3 and video conferencing.

The current deployment of 1xEV-DO systems is referred to as the Revision 0 system [Rev0]. The newly standardized version of 1xEV-DO system is referred to as the Revision A (Rev A) system [RevA]. Many improvements have been incorporated in the Rev A system to support multimedia applications including VoIP with better quality of service (QoS) control. The improvements include enhanced reverse link throughput via HARQ and rate adaptation, flexible rate adaptation and multipackets multiplexing on the forward link, and enhanced scheduling and rate control. The 1xEV-DO Rev A system will be deployed in 2006.

1xEV-DV provides integrated voice and high-speed data services. It delivers a peak data rate of 3.09 Mbps in the forward direction, with enhanced modulation and coding technique. It is designed to optimize the resource allocations for both voice and high-speed data services. It explores power control for voice and low-speed data services and rate control for high-speed packet data services. The 1xEV-DV solution standards Revision D were finalized at the beginning of 2004 [RevD_1]-[RevD_4].

CDMA2000 has benefited from the extensive experience acquired through multiple years of operation of the cdmaOne systems. As a result, CDMA2000 is a very efficient and mature technology and has been tested in various spectrums. The commercial success of CDMA2000 has made the 3G vision a reality. It currently dominates the 3G markets in North America today and will continue to lead in the future.

4.2   Overview of CDMA2000 3G1x Network Architecture

Figure 4-2 shows the high-level network architecture of a CDMA2000 3G1x system. The network includes two parts: radio access network (RAN) and core network (CN). The radio access network consists of the following major functional components:
- 3G1x voice or packet data mobiles, i.e., mobile handset or laptop computer.

- Base station Transmission System (BTS)
- Radio Network Controller (RNC)
- Mobile Switch Center (MSC)
- Packet Control Function (PCF)
- Network Element Management System (EMS)
- Communication link between the BTS and RNC or MSC, which is called the backhaul.

The core network consists of the following major functional components:
- Packet Data Service Node (PDSN)
- Public Switch Telephone Network (PSTN)
- Public/Private Packet Data network
- Authentication, Authorization, Accounting (AAA) Server

The CDMA2000 system supports both 2G and 3G handsets, as well as laptop computers or personal data assistant (PDA) with 3G high-speed data capabilities.

The BTS performs the function of communicating to the access terminal (AT) over the air interface. It contains the software and hardware to perform the digital communications and signal processing that is required to implement the CDMA2000 air interface technologies. It also contains the Radio Frequency (RF) components that are required to transmit and receive RF signals over the air to and from the AT. The BTS communicates with the RNC and MSC through the backhaul links. Depending on the implementation, back-end servers and routers may exist in the backhaul to facilitate the communication between the BTS and RNC or MSC.

The RNC provides signaling and traffic processing control for each data session. These functions include session establishment and release, frame selection and Radio Link Protocol (RLP) processing, and signaling processing. The RNC acts as a central controlling entity that manages the signaling and traffic routed to each BTS for mainly data applications.

The MSC is the control and switch center that manages the signaling and traffic processing for voice calls. It provides the main call processing functions such as call setup and tear-down, paging, registration, frame selection for soft handoff calls, etc. It also has the voice coder that translates other voice coding formats to the formats defined by the CDMA2000 specifications. The MSC interfaces to other MSCs to establish the mobile-to-mobile calls or to the PSTN to establish the mobile to wire-line phone calls.

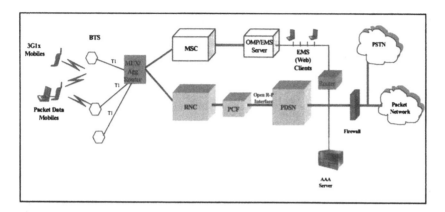

Figure 4-2. 3G1x Network Architecture

It interfaces to the EMS servers and billing system to support all the operation and management functions.

The PCF provides the processing for the standard A10/A11 interface to the PDSN and allows the RNC functions to interface to the PDSN. The A10/A11 interface terminates all mobility management functions of the radio network, and is the demarcation point between the RAN and the Packet Data Network.

The EMS provides operation, administration, maintenance, and provisioning functions for the network. It interacts with the RAN network components via protocols such as SNMP.

The PDSN resides in the serving network and is allocated by the serving network where a mobile terminal (MT) initiates a service session. It terminates the PPP link protocol with the MT. The PDSN maintains link layer information with the PCF, and routes packets to external packet data networks or to the HA (Home Agent) in the case of tunneling to the HA. The PDSN acts as a gateway interfacing between the RAN and external packet data networks, connecting to the PCF by the A10/A11 interface and maintaining an interface to the backbone packet data network.

The AAA server provides the authentication, authorization, and accounting functions of the network. The database contains service and protocol information for each user registered in the network. This information is returned to the PDSN using a secure protocol when the user first establishes a session in the network.

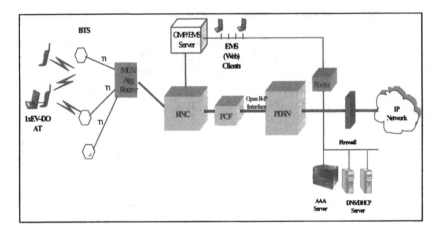

Figure 4-3. 1XEV-DO Network Architecture

4.3   Overview of CDMA2000 1xEV-DO Network Architecture

The network architecture of 1xEV-DO is very similar to the 3G1x network architecture except that the components associated with voice traffic are absent. Figure 4-3 shows the high-level reference architecture for a 1xEV-DO system. Since the 1xEV-DO system supports high-speed data users only, the MSC in the 3G1x network architecture is not needed anymore. The mobile terminal is usually called the Access Terminal (AT) in 1xEV-DO systems. Other components remain the same, providing similar functions specific to the 1xEV-DO services.

The 1xEV-DO RAN is capable of supporting the Internet Protocol (IP) for packet data users. Additional IP components are used to facilitate the management of the IP networks, such as the DNS (Domain Name Service) and DHCP (Dynamic Host Configuration Protocol) server.

The backhaul transport in 1xEV-DO is IP based instead of Frame Relay in 3G1x or ATM in UMTS network.

Depending on implementation, 1xEV-DO system can support simple IP, Mobile IP or other layer 2 tunneling protocol such as L2TP for network layer communications. For mobile IP configuration, HA (home agent) and

FA (foreign agent) nodes are needed. PDSN usually functions as a foreign agent. The end-to-end protocol stacks vary with different configurations.

4.4   Overview of 3GPP Standards Evolution

In Europe, the standardization organizations involved in the creation of the Third-Generation Partnership Project 3GPP were ETSI (Europe), ARIB (Japan), TTA (Korea), TTC (Japan), and T1P1 (USA). Later during 1999, CWTS (the China Wireless Telecommunication Standard Group) also joined 3GPP. The basic third-generation research work was started under the RACE I (Research of Advanced Communication Technologies in Europe) program in 1988. This program was followed by RACE II, which did research in CDMA-based CODIT as well as TDMA-based ATDMA air interface between 1992 and 1995. Then, the Advanced Communication Technologies and Services (ACTS) program was launched at the end of 1995 to support mobile communications research and development. Within ACTS, the Future Radio Wideband Multiple Access System (FRAMES) project [Nik98] was setup to define a proposal for a UMTS radio access system. A harmonized multiple access platform was defined, consisting of two modes: FMA1, a wideband TDMA, and FMA2, a wideband CDMA. The wideband TDMA and the FRAMES wideband CDMA proposals were submitted to ETSI as candidates for UMTS air interface.

Within 3GPP, four different technical specification groups (TSG) were set up:
- Service and System Aspects TSG
  This TSG defines UMTS services, overall system architecture, etc.
- Terminals TSG

    This TSG defines all the protocols between the Cu/Uu interface and functions that need to be supported within the terminals. The Cu interface is between the mobile equipment (ME) and the subscriber identity module (SIM). The Uu interface is between the ME and the base station.

- Radio Access Network TSG
  This TSG defines the UTRAN network architecture and all the protocols related to the RAN, e.g., the various protocols for the interfaces between the Radio Network Controller and Node B, etc.

- Core Network TSG

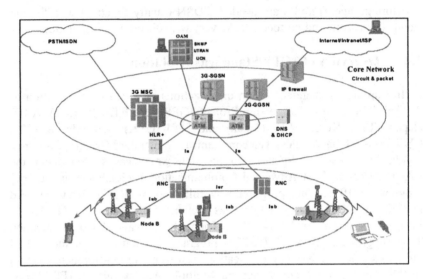

Figure 4-4. UMTS R99/4 Network Architecture

This TSG defines the core network architecture and all the associated protocols, e.g., mobility management, session management, etc.

4.5   Overview of UMTS R99/4 Network Architecture

The high-level UMTS system architecture is shown in Figure 4-4. It consists of (1) a Radio Access Network (RAN) or UMTS Terrestrial RAN (UTRAN) that handles all radio-related functionalities; (2) a Core Network (CN), which is responsible for switching calls and routing data packets to external network; and (3) User Equipment (UE) that interfaces with the UMTS network.

Completely new protocols based on the new WCDMA radio technology are designed for both the UE and UTRAN but most of the CN is inherited from the GSM/GPRS networks.

The UE consists of two parts:
- The UMTS Subscriber Identity Module (USIM) is a smartcard that stores the subscriber identity, authentication and encryption keys, and some subscription information that is required at the terminal. It also performs authentication algorithms.

- The Mobile Equipment (ME), which is the radio terminal used for radio communications over the Uu interface.

UTRAN consists of a set of Radio Network Subsystem (RNS) connected to the CN via the Iu interface. A RNS consists of a Radio Network Controller (RNC) and one or more NodeBs. A NodeB is connected to the RNC via the Iub interface. The two network elements in UTRAN are:

- NodeB, which controls the data flow between the Uu and Iub interfaces. It terminates the physical layer, extracts the MAC protocol data units, and transports them across the Iub interface to the RNC. It also participates in radio resource management.
- RNC, which controls the radio resources in its domain. RNC is the service access point for all services UTRAN provides to the CN.

The main elements of the CN are:

- Home Location Register (HLR): a database located in the user's home system. It stores the master copy of the user's service profile. A service profile consists of information on allowed services and supplementary service information, e.g., the call forwarding number, forbidden roaming areas, etc. A service profile is created when a new user subscribes to the system and remains stored as long as the subscription is active. In addition, the HLR also stores the UE location on the level of MSC/VLR and/or SGSN so that incoming transactions (e.g., calls) can be routed to the UE.

- Mobile Services Switching Center/Visitor Location Register (MSC/VLR): MSC switches the Circuit Switched (CS) calls while the VLR contains more precise information on the UE's location and a copy of the visiting user's service profile.

- GMSC (Gateway MSC) is the switching point where UMTS Public Land Mobile Network (PLMN) is connected to the external CS networks. All incoming and outgoing CS connections go through GMSC.

- Serving GPRS Support Node, SGSN's functionality is similar to that of MSC/VLR but is used for Packet Switched (PS) services.

- Gateway GPRS Support Node (GGSN) is equivalent to a GMSC in the PS domain.

The external networks can be divided into two groups:

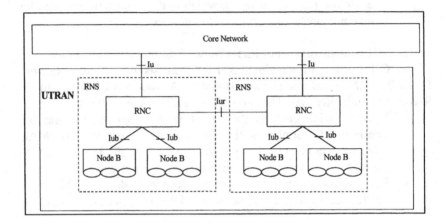

Figure 4-5. UTRAN Architecture [25.401]

- CS networks, which provide circuit-switched connections, e.g., ISDN and PSTN.

- PS networks, which provide connections for packet data service, e.g., the Internet.

The UMTS standards define the interfaces between the logical network elements. The main open interfaces defined are:
- Cu Interface: the interface between the USIM smartcard and the ME
- Uu Interface: the WCDMA radio interface through which the UE assesses the fixed part of the system
- Iub Interface: the interface that connects a Node B and an RNC.
- Iur Interface: the interface between different RNCs
- Iu Interface: the interface that connects UTRAN to the CN. It has a CS and a PS portion. The CS portion is referred to as Iu-CS and the PS portion is referred to as Iu-PS.

4.5.1 UTRAN Components

UTRAN (as shown in Figure 4-5) consists of two types of network elements: NodeB and RNC.

4.5.1.1 The NodeB

The main function of the NodeB is to perform the air interface layer 1 processing (channel coding and interleaving, rate adaptation, spreading, etc.)

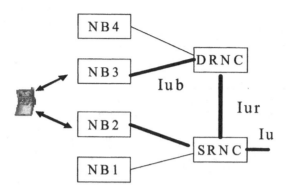

Figure 4-6. Data Flow between SRNC/DRNC

and some basic radio resource management operations, e.g., the inner loop power control.

The RNC is the network element responsible for the control of the radio resources of UTRAN. It terminates the Radio Resource Control (RRC) protocol, which defines the messages and procedures between the mobile and UTRAN. The RNC connects to the CN normally via one MSC and one SGSN.

The RNC, which controls a NodeB (terminating the Iub interface toward the NodeB), is referred to as the Controlling RNC (CRNC) of the NodeB. The CRNC is responsible for controlling the loading of its own cells by performing the admission and congestion control of its own cells. It also performs code allocation for newly established radio links in these cells.

Sometimes, a mobile UTRAN connection uses resources from more than one Radio Network System (RNS). In such cases, the RNCs involved have two separate logical roles (referred to Figure 4-6):

- Serving RNC (SRNC). Each Mobile Equipment (ME) connected to UTRAN has one and only one SRNC. A SRNC for a mobile performs the layer 2 processing of the data to/from the radio interface. It also performs some basic radio resource management operations, e.g., outer loop power control, the handover decisions, and the mapping of the Radio Access Bearer parameters into the air interface transport channel parameters. The SRNC also terminates both the Iu link for the transport of user data and the corresponding

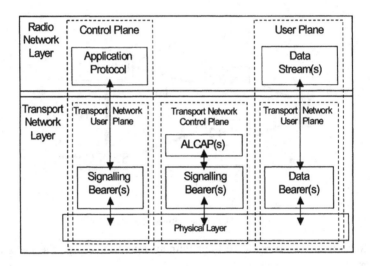

Figure 4-7. General Protocol Model for Utran Interfaces [25.401]

RANAP signaling to/from the core network. In addition, it terminates the radio resource controlling signaling, i.e., the signaling protocol between the Mobile Equipment (ME) and UTRAN. The SRNC may also be the CRNC of some NodeBs used by the mobile in an existing connection with the UTRAN.

- Drift RNC. Each UE may have zero, one, or more DRNCs. A DRNC is any RNC, other than the SRNC, that controls cells used by the mobile. The DRNC routes the data between the Iub and Iur interfaces, except when the UE is using a common or shared transport channel. It does not perform layer 2 processing of the user plane data.

4.5.2   General Protocol Model for UTRAN Terrestrial Interfaces

The protocol structures in UTRAN interfaces are designed according to a general protocol model as shown in Figure 4-7 [25.401]. The basic principle for the general protocol model is that the user planes and the control planes should be logically independent of each other.

Horizontally, the protocol structure consists of two main layers: the radio network layer (RNL) and the transport network layer (TNL). All UTRAN-

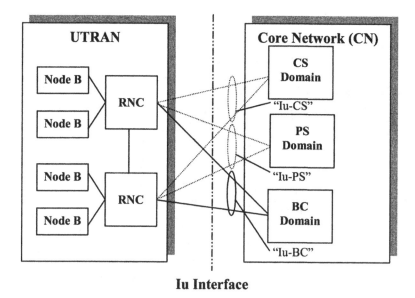

Iu Interface

Figure 4-8. Iu Interface Architecture [25.410]

related issues are visible only in the RNL. The TNL represents a standard transport technology that is selected to be used in UTRAN.

Vertically, we have the control plane, the user plane, and the transport network control and user planes. The control plane is used for all UMTS-specific controlling signaling. It includes the Application Protocol, i.e., RANAP (for Iu), NBAP (for Iub), and RNSAP (for Iur), and the Signaling Bearer for transporting the Application Protocol messages. The Application Protocol is used for setting up the radio access bearers in Iu and the Radio Link in Iub and Iur to the UE. ALCAP is used to set up the data bearer within the transport network user plane. The Signaling Bearer for the Application Protocol may or may not be of the same type as the Signaling Bearer for the ALCAP. It is set up via OA&M operations.

The user plane transports all information sent and received by the user. The user plane includes the data streams and the data bearers for the data streams. Each data stream is characterized by one or more frame protocols specified for that interface.

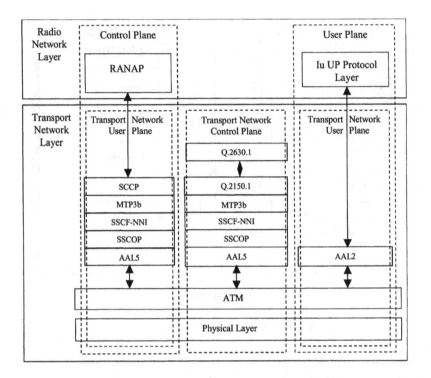

Figure 4-9. Iu-CS Protocol Structure [25.410]

The transport network control plane is used for all control signaling within the transport layer. It includes the ALCAP protocol that is needed to set up transport bearers (Data Bearer) for the User Plane. It also includes the Signaling Bearer needed for the ALCAP.

The transport network user plane includes the Data Bearers in the User Plane and the Signaling Bearers for the Application Protocol.

4.5.2.1 Protocol Structures for the UTRAN-CN Interface Iu

The Iu interface connects UTRAN to CN [25.410] as shown in Figure 4-8. It has two different instances: Iu CS (Iu Circuit Switching) for connecting UTRAN to Circuit Switched CN, and Iu PS (Iu Packet Switching) for connecting UTRAN to Packet Switched CN.

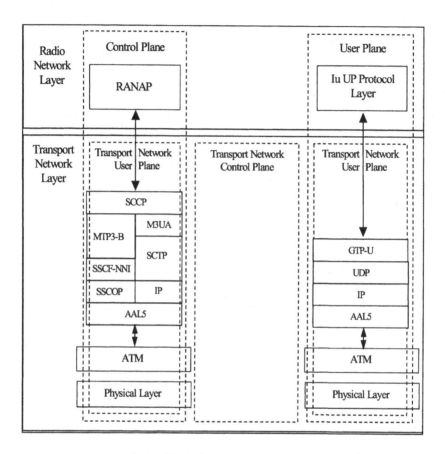

Figure 4-10. Iu PS Protocol Structure [25.410]

The overall Iu-CS protocol structure is shown in Figure 4-9. All three planes in the Iu interface share a common ATM transport. The Iu-CS control plane protocol stack consists of RANAP, on top of broadband SS7 protocols. The applicable layers are the Signaling Connection Control Part (SCCP), the Message Transfer Part (MTP3-b) and Signaling ATM Adaptation Layer for Network-to-Network Interfaces (SAAL-NN1). SAAL-NNI is further divided into Service Specific Co-ordination Function (SSCF), Service Specific Connection Oriented Protocol (SSCOP), and ATM Adaptation Layer 5 (AAL) layers. SSCF and SSCOP layers are specifically designed for managing signaling connections in ATM networks.

The Iu-CS Transport Network Control Plane protocol stack consists of the Signaling Protocol for setting up the AAL2 Connection (Q.2630.1 and

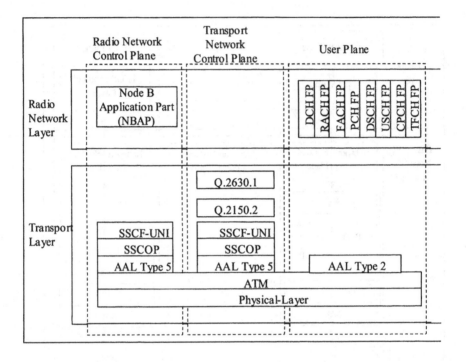

Figure 4-11. Iub Protocol Structure [25.430]

adaptation layer Q.2150.1), on top of BB SS7 protocols. A dedicated AAL2 connection is reserved for each individual CS service.

The Iu-PS protocol structure is shown in Figure 4-10. A common ATM transport can be used for both User and Control Plane. The Control Plane protocol stack again consists of RANAP, and the same Broadband SS7-based signaling bearer. An alternative IP-based signaling bearer is specified. The SCCP layer is used commonly for both Broadband SS7-based or IP-based signaling bearer. The IP-based signaling bearer consists of M3UA (SS7 MTP3 User Adaptation Layer), SCTP, and IP. AAL5 is common to both alternatives. The SCTP layer is specifically designed for signaling transport in the Internet. The transport network control plane does not apply to Iu-PS. In the Iu-PS User Plane, multiple GTP tunnels are multiplexed on one or several AAL5 PVCs. The User Plane of the GPRS Tunneling Protocol (GTP-U) is the multiplexing layer that provides identities for individual packet data flow. Each flow uses UDP connectionless transport and IP addressing. Setting up of the GTP tunnel requires only an identifier for the tunnel, and the IP addresses for both directions, and these parameters are already included in the RANAP RAB Assignment messages.

4.5.2.2 Protocol Structures for Iub, the RNC–NodeB Interface

The protocol stack of the RNC-NodeB interface (Iub interface) is shown in Figure 4-11. The Iub interface signaling, Node B Application Part (NBAP), is divided into two essential components: the common NBAP, which defines the signaling procedures across the common signaling link, and the dedicated NBAP, used in the dedicated signaling link. The main functions of the Common NBAP are:

- Cell configuration management
- Resource event management
- Setup, reconfiguration, and deletion of common transport channels
- Radio link setup
- Common measurement initialization, reporting and terminations
- Fault management

The main functions of the dedicated NBAP are:
- Handling of dedicated channels
- Addition, reconfiguration, and deletion of radio links for one UE context
- Dedicated measurement initialization, reporting, and termination
- Downlink power control
- Dedicated radio link fault management

The User Plane Iub frame protocols define the basic inband control procedures for every type of transport channel and the structures of the frames. The Q.2630.1 signaling is used for the dynamic management of the AAL2 connections used in the User Plane.

4.5.2.3 Protocol structures for Iur, the RNC-RNC Interface

The protocol stack of the RNC-to-RNC interface, i.e., Iur interface, is shown in Figure 4-12. The Iur interface provides four distinct functions:

- Support of the basic Inter-RNC mobility

This requires the basic module of RNSAP signaling as described in [25.423]. It provides the support for the mobility of the user between the two RNCs but does not support the exchange of any user data traffic.

The functions offered by the Basic Inter-RNC mobility module include the support of SRNC relocation, e.g., uplink/downlink signaling transfer, inter-RNC packet paging, inter-RNC cell and UTRAN registration area update, and the reporting of protocol errors.

- Support of Dedicated Channel (DCH) Traffic

This functionality requires the dedicated channel module of the RNSAP signaling and allows the transport of dedicated channel traffic between two RNCs. This functionality also requires support of the User Plane Frame Protocol for the dedicated channel plus the Transport Network Control Plane Protocol (Q.2630.1), which is used for the setting up of the transport connections (AAL2 connections). The Frame Protocol for dedicated channels, in short DCH FP [25.427], defines the structure of the data frames carrying the user data and the control frames used to exchange measurements and control information.

The functions offered by the Iur DCH module are:
- o Setup and release of dedicated transport connections across the Iur interface
- o Transfer of DCH transport blocks between SRNC and DRNC
- o Establishment, modification (e.g., via radio link reconfiguration and physical channel reconfiguration procedures), and release of the dedicated channel in the DRNC due to hard and soft handover
- o Management of the radio links in the DRNS, via dedicated measurement report procedures and power control procedures

- Support of Common Channel (CCH) Traffic

This functionality allows the handling of common and shared channel data streams across the Iur interface. It requires the Common transport Channel module of the RNSAP protocol and the Iur Common transport Channel Frame Protocol (CCH FP). The Q.2630.1 signaling protocol of the Transport Network Control Plane is also needed if signaled AAL2 connections are used.

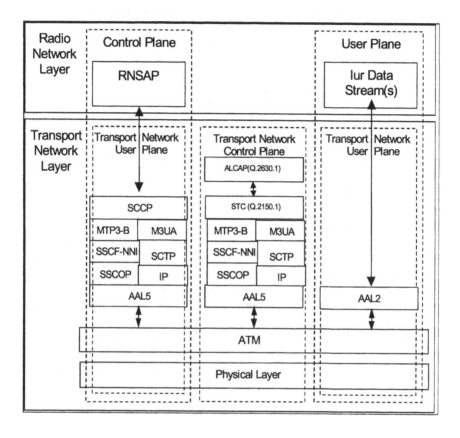

Figure 4-12. Iur Protocol Structure [25.420]

The functions offered by the Iur common transport channel module are:

- Setup and release of the transport connection across the Iur for common channel data streams
- Splitting of the MAC layer between the SRNC (MAC-d) and the DRNC (MAC-c and MAC-sh). The scheduling for Downlink data transmission is performed in the DRNC
- Flow control between the MAC-d and MAC-c/MAC-sh
- Support of Global Resource Management

This functionality provides signaling to support enhanced radio resource and OAM features across the Iur interface. It is implemented via the global module of the RNSAP protocol, and does not require the User Plane protocol. The functions offered by the Iur global resource module are the transfer of cell measurements between two RNCs and the transfer of NodeB timing information between two RNCs.

4.5.3   Core Network Components

The three main logical elements in a UMTS Core Network are the MSC (for circuit voice/data services), Serving GPRS Support Node (SGSN), and Gateway GPRS Support Node (GGSN) for packet data services.

4.5.3.1 The MSC

The MSC architecture (shown in Figure 4-13) usually consists of multiple components, namely a feature server and multiple access gateways. The feature server performs mobility management and call processing, and terminates signaling protocols and control bearer gateways. One access gateway is the wireless access gateway that terminates the protocol interface from the radio access network, performs voice coding and provides tones. Another access gateway is the trunk access gateway that provides the trunk interface to PSTN and other MSCs. The multimedia resource server (MRS) provides announcements and conferencing. All bearer traffic between the wireless access gateway, trunk access gateway, and MRS is carried over an internal IP network using RTP/UDP transport.

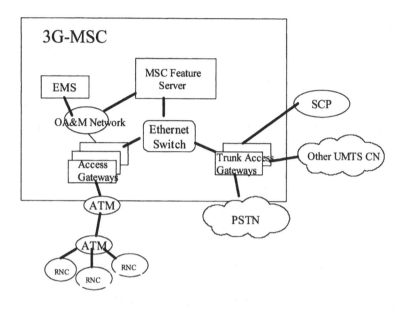

Figure 4-13. MSC Architecture

4.5.3.2 The Serving GPRS Support Node (SGSN)

A SGSN transports RAN-side control messages and end-user traffic from UTRAN to the core network via ATM PVCs. The RAN-side control messages and end user traffic may be carried on separate ATM PVCs. The RAN-side control messages include messages between the RNCs and the SGSN and between the mobile and SGSN. All RAN-side control messages are carried within a protocol layer called RAN Application Part (RANAP). RANAP supports the general functions of Packet Mobility Management (PMM), Session Management (SM), and Short Message Service (SMS). PMM includes mobile attach, and HLR interactions for authentication and service authorization and routing area tracking. SM supports creating and maintaining packet data sessions each of which is referred to as a Packet Data Protocol (PDP) context. RANAP is carried over SCCP over Broadband SS7 (SCCP/MTP-3B-SAAL/ATM).

End-user traffic between the RNC and SGSN is carried over the GPRS Tunneling Protocol User Part (GTP-U). User packets are encapsulated within GTP-U tunnels between the RNC and SGSN. GTP is also used

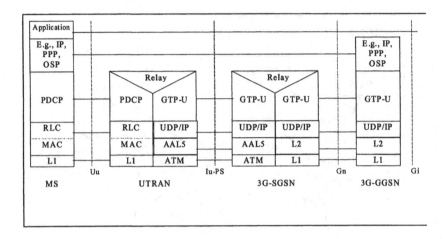

Figure 4-14. Protocol Stacks for User Plane for Release 99 [23.060]

between the SGSN and GGSN. GTP-U on the RAN side is carried over the UDP/IP/AAL5ATM. GTP is described in 3GPP TS29.060 [29.060].

The SGSN has narrowband SS7 interfaces to support signaling with HLRs, MSCs, SMS-SCs, and SCPs. For all these end points, the lower layers of the stacks are SCCP(connectionless)/MTP3/MTP2. For signaling with HLRs and SMS-SCs, the application layer is Mobile Application Part (MAP) over TCAP. For communication with MSCs, the application layer is BSSAP+. The application layer for CAMEL is called CAMEL Application Part (CAP).

The SGSN is analogous to an MSC in that it communicates with an HLR to register a mobile and obtain a profile of authorized services, performs authentication and service authorization, creates and maintains packet data connections, and collects charging data.

4.5.3.3 The Gateway GPRS Support Node (GGSN)

The GGSN is the gateway router that connects the UMTS packet data users to the external network. GGSN also performs mobility management, and relays packets destined to the registered mobiles to relevant SGSNs via GTP tunnels. If necessary, it can perform address translation and mapping functions, data compression, and encryptions. GGSN also provides message screening function and collects charging data.

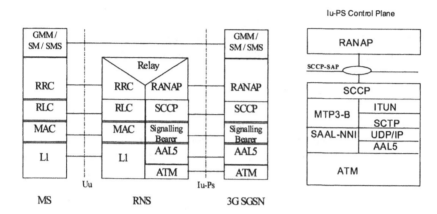

Figure 4-15. Protocol Stacks for Control Plane for Release 99 [23.060]

The interface between the SGSN and GGSN is the GPRS Tunneling Protocol. It is separated into a control plane (GTP-C) and a user plane (GTP-U). GTP-C is used to set up, maintain, and tear down user data connections while the user data is sent over GTP-U. All GTP-C and GTP-U messages are transported over UDP/IP [29.060].

4.5.4   General Protocol Model for CN Interfaces

The protocol stacks between the logical elements of a R99 UMTS network for user and control planes are shown in Figures 4-14 and Figure 4-15, respectively.

The packet data network (PDN) is an IP network providing connectivity from the corresponding host (CH) to the gateway GSN (GGSN). Between GGSN and the serving GSN (SGSN), IP packets are transported via the GRPS tunneling protocol (GTP) which is used for both data and signaling. The network connecting the GSNs within a PLMN and between PLMNs is a private IP network. In the case of IP packets encapsulated by GTP, UDP is used to carry the GTP PDU. At the SGSN, the original IP packet is recovered and encapsulated again using GTP for transporting to the UTRAN. At RNC, the packet is extracted and segmented into appropriate RLC/MAC protocol data units. The radio link control (RLC) between the RNC and the MS provides a highly reliable connection.

For the control plane, the Radio Resource Control (RRC) connection is established between Mobile Station (MS) and the Radio Network Subsystem (RNS). The session management (SM), group mobility management (GMM), and SMS connection is established between MS and 3G-SGSN. RANAP is established between the RNS and the 3G-SGSN.

4.6   Mobility Management

There are two basic operational modes of a UE: idle and connected mode. The connected mode is further divided into different service states depending on the kind of physical channels a UE is using, e.g., Cell_DCH, Cell_FACH, Cell_PCH, and URA_PCH states as shown in Figure 4-16.

In the idle mode, after the UE is switched on, it selects a Public Land Mobile Network (PLMN) to contact. The UE looks for a suitable cell of the chosen PLMN to provide services and tunes to its control channel. This process is known as "camping on a cell." The UE stays in idle mode until it transmits a request to establish an RRC connection. In idle mode, a UE is identified by non-access stratum identities such as IMSI, TMSI, and P-TMSI.

In the Cell_DCH state, a dedicated physical channel is allocated to the UE and the UE is known by its serving RNC on a cell or active set level. The UE performs measurements and sends measurement reports according to the measurement information received from the RNC.

In the Cell_FACH state, no dedicated physical channel is allocated for the UE but the UE can use the RACH/FACH channels to transmit both signaling messages and small amount of user plane data. In this state, the UE is capable of receiving system information from the broadcast channel. The UE also performs cell reselection and sends a cell update to the RNC upon cell reselection so that the RNC knows the UE location on a cell level.

In the URA_PCH state, the UE does not execute a cell update after each cell reselection, but instead reads UTRAN registration area (URA) identities from the broadcast channel, and only if the URA changes (after cell reselection) does UE pass its location to the SRNC using the URA update procedure.

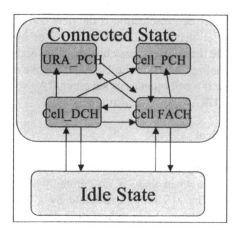

Figure 4-16. RRC Modes and Connection State

In the Cell_PCH state, the UE listens to the Broadcast channel and performs cell updates. As in the URA_PCH state, the UE is not allowed to use any uplink channel.

4.6.1 Circuit-Switched Services

When the UE is in the Cell_DCH state, a normal way to perform mobility control is to use Active Set Update and Hard Handover procedures. The purpose of the Active Set Update is to update the active set of the connection between the UE and the UTRAN while the UE is in the Cell_DCH state. The handover algorithm for UMTS is described in [25.922]. The Hard Handover procedure can be used to change the radio frequency band of the connection between the UE and UTRAN or to change the cell on the same frequency when no network support of macro diversity exists. Intersystem handover procedure can be used for UE in either Cell_DCH or Cell_FACH states to handover a connection from UTRAN to another radio access system, e.g., GSM and vice versa.

When the UE is in the Cell_FACH/Cell_PCH/URA_PCH state, specific procedures are used to keep track of the UE location either on a cell level or on a URA level. A Cell Update procedure is used by a UE in the Cell_FACH/Cell_PCH states. It can be triggered for several reasons, e.g., cell reselection, expiration of periodic cell update timer, initiation of uplink data transmission, or UTRAN-originated paging.

The URA Update procedure is used by UE in the URA-PCH state. It can be triggered either after cell reselection if the new cell does not contain the same UTRA identifier that the UE currently has or by expiration of the periodic URA Update timer. A cell may broadcast several URA identifiers to avoid excessive signaling. A UE in the URA-PCH state always has one and only one valid URA. If a cell broadcasts several URAs, the RNC assigns one URA to a UE in the URA Update confirm message.

4.6.2   Packet Services

There are two mobility management states for packet services. A user is in PMM-detached state when there is no communication between the MS and the 3G-SGSN. The MS is not reachable by a 3G-SGSN as the MS location is not known. When the mobile station performs a GPRS attach, a PS connection is set up and the mobility state changes to PMM-connected in the 3G-SGSN and in the MS. The PS connection is made up of two parts: an RRC connection and an Iu connection. In the PMM-connected state, the location of the MS is tracked by the serving RNC. When the PS signaling connection is released, the mobility state changes to PMM-idle. When the MS performs a GPRS detach, the mobility state changes from PMM-idle to PMM-detached. Similarly, if the GPRS attach is rejected or the MS performs a GPRS detach while in PMM-connected state, the mobility state will also change to PMM-detached.

A routing area update takes place when a GPRS-attached MS detects that it has entered a new Routing Area, when the RA update timer has expired, or when the MS has new access capabilities to indicate to the network. When the SGSN, which receives the routing area update, realizes that this is the first time it has seen the MS, it will send a SGSN context request to the old SGSN to get the mobility management/PDP contexts for the MS. After authenticating the user, it will send an update PDP context request to the GGSN as well as an update location to the HLR. Only when all such procedures are successful will the SGSN send a routing area update response to the MS.

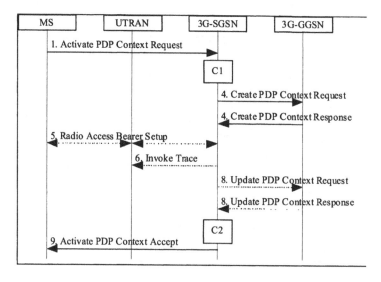

Figure 4-17. PDP Context Activation [23.060]

After executing a GPRS attach, a MS may initiate an activate PDP context procedure to start a packet data session. The PDP context activation procedure is shown in Figure 4-17. For packet data service, the SGSN performs both the mutual authentication and security keys agreement. The SGSN obtains the authentication vector from HLR and stores it. The SGSN then performs the challenge and response steps with the mobile. User data are transferred transparently between the MS and the external data networks using encapsulation and tunneling. GPRS Tunneling Protocol (GTP) [29.060] is used to tunnel user data from an RNC to a SGSN and later from a SGSN to a GGSN.

In Figure 4-18, we consider the case of an MS roaming away from its home public land mobile network (PLMN) in a visiting PLMN. The GSN connected to the MS is called the serving GSN (SGSN). SGSN can access the visiting location register (VLR) which is located in a mobile switching center (MSC). However, the MS is registered at the home location register (HLR) which can be accessed by the gateway GSN (GGSN). A corresponding host (CH) in the packet data network (PDN) would send the IP packet to the MS through the GGSN first. Such triangular routing is the cause of large end-to-end delay for UMTS packet data service. An improved version in which a local GGSN within the visiting network is chosen for the roaming mobile so that triangular routing can be eliminated has been

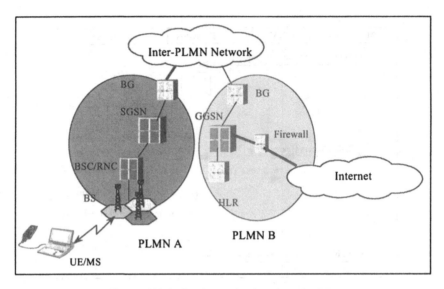

Figure 4-18. Packet Connection for a Roaming UE

discussed in 3GPP. For more elaborate description of the mobility management procedures for UMTS packet data service, readers can refer to [23.060].

Acknowledgments

The authors would like to thank 3GPP for giving permission to reproduce some of the diagrams from the various 3GPP TSG standard documents. 3GPP TSs and TRs are the property of ARIB, ATIS, ETSI, CCSA, TTA, and TTC, who jointly own the copyright for them. They are subject to further modifications and are therefore provided to you "as is" for information purposes only. Further use is strictly prohibited.

4.7   Review Exercises

1. Name the two major networks that make up a UMTS system and the major network elements within each network.
2. List the roles of Serving RNC and Drift RNC when a mobile uses resources from more than one RNS
3. Explain how mobility management is supported for Circuit Switched and Packet Switched Services in UMTS.

4.8   References

[Rev0] 3rd Generation Partnership Project 2 (3GPP2), "cdma2000 High Rate Packet Data Air Interface Specification," Technical Report C.S20024-0 v2.0, October 2000.

[RevA] 3GPP2, "cdma2000 High Rate Packet Data Air Interface Specification," Technical Report C.S20024-A v1.0, April 2004.

[RevD_1] 3GPP2, "Introduction to cdma2000 Standards for Spread Spectrum Systems, C.S0001-D, February 2004.

[RevD_2] 3GPP2, "Physical Layer Standard for cdma2000 Spread Spectrum Systems," C.S0002-D, February 2004.

[RevD_3] 3GPP2, "Medium Access Control (MAC) Standard for cdma2000 Spread Spectrum Systems," C.S0003-D, February 2004.

[RevD_4] 3GPP2, "Signaling Link Access Control (LAC) Standard for cdma2000 Spread Spectrum Systems," C.S0004-D, February 2004.

[Nik98] E. Nikula et al, "FRAMES Multiple Access for UMTS and IMT-2000," IEEE Personal Communications Magazine, April 1998, pp 16-24.

[23.060] 3GPP Technical Specifications, 23.060 GPRS Service Description Stage 2.

[25.401] 3GPP Technical Specifications, 25.401 UTRAN Overall Description.

[25.410] 3GPP Technical Specifications, 25.410 UTRAN Iu Interface: General Aspects and Principles

[25.412] 3GPP Technical Specifications, 25.412 UTRAN Iu Interface: Signalling Transport

[25.413] 3GPP Technical Specifications, 25.413 UTRAN Iu Interface: RANAP Signaling.

[25.414] 3GPP Technical Specifications, 25.414 UTRAN Iu Interface: Data transport and Transport Signaling.

[25.420] 3GPP Technical Specifications, 25.420 UTRAN Iur Interface: General Aspects and Principles

[25.422] 3GPP Technical Specifications, 25.422 UTRAN Iur Interface: Signalling Transport

[25.423] 3GPP Technical Specifications, 25.423 UTRAN Iur Interface: RNSAP Signaling.

[25.424] 3GPP Technical Specifications, 25.424 UTRAN Iur Interface: Data transport and Transport Signaling for CCH Data Streams.

[25.425] 3GPP Technical Specifications, 25.425 UTRAN Iur Interface: User Plane Protocols for CCH Data Streams.

[25.426] 3GPP Technical Specifications, 25.426 UTRAN Iub/Iur Interface: Data transport and Transport Signaling for DCH Data Streams.

[25.427] 3GPP Technical Specifications, 25.427 UTRAN Iub/Iur Interface: User Plane Protocols for DCH Data Streams.

[25.430] 3GPP Technical Specifications, 25.430 UTRAN Iub Interface: General Aspects and Principles

[25.432] 3GPP Technical Specifications, 25.432 UTRAN Iub Interface: Signaling Transport

[25.433] 3GPP Technical Specifications, 25.433 UTRAN Iub Interface: NBAP Signaling.

[25.922] 3GPP Technical Specifications, 25.922 Radio Resource Management Strategies.

[29.060] 3GPP Technical Specifications, 29.060 GPRS Tunneling Protocol.

Chapter 5

AIR INTERFACE PERFORMANCE AND CAPACITY ANALYSIS

5. CAPACITY ANALYSIS AND EVALUATION

5.1   Queuing Analysis in a Wireless Communication System

For conventional frequency division and time division multiple access techniques, Erlang capacity is widely used to measure the average number of users who can be served at a time with a certain quality. In these systems, a slot, either in frequency and/or time domain, is assigned to each user to provide services. Queuing analysis [Gro98] is performed to establish the relationship between the number of users and the available slots. The capacity is a function of the probability that slots are available to the user who requests service at a given time. Queuing theory was first introduced by Erlang [Ber87] for wireline telephone traffic. *Erlang* now is a unit of measurement of telephone traffic. It is equal to one hour of conversation, i.e., 3,600 seconds or 36 hundred call-seconds (CCS).

In a CDMA system with spread spectrum multiple access technique, all users share the same spectrum and bandwidth. There is no distinguishing notion of slots such as those in FDMA and TDMA systems. Other issues such as user mobility, handoff, and data rate per user, make it difficult to analyze and characterize the traffic in a mobile system. Nevertheless, the traditional queuing analysis is applied with variants to facilitate the system planning and design.

5.1.1   Call Arrival Process

Typically, the call arrival is modeled as a Poisson process [Pap91]. In a time interval of t, the probability of k arrivals is given by

$$p_k = \frac{(\lambda t)^k}{k!} e^{-\lambda t} \qquad (5\text{-}1)$$

where λ is the average arrival rate (calls per second). Equation (5-1) is the Poisson probability distribution. Its mean and variance are the same, i.e., $E[k] = \lambda t$, $E[k^2] - E^2[k] = \lambda t$.

5.1.1.1 Interarrival time

Let τ denote the time between adjacent calls, which is a random variable. The probability distribution function of τ is given by

$$F_\tau(t) = \Pr\{\tau \le t\} = 1 - \Pr\{\tau > t\} \qquad (5\text{-}2)$$

However $\Pr\{\tau > t\}$ is exactly the probability that there is no arrival in the t interval, which is p_0 in equation (5-1). Thus we have

$$F_\tau(t) = 1 - e^{-\lambda t} \qquad (5\text{-}3)$$

Differentiating equation (5-3) with respect to t, we obtain the density function as

$$f_\tau(t) = \lambda e^{-\lambda t} \qquad (5\text{-}4)$$

The mean interarrival time is thus $E[t] = 1/\lambda$.

Therefore, for a Poisson arrival, the interarrival time is exponentially distributed with mean of $1/\lambda$.

5.1.1.2 Call Holding Time

The call holding time is typically assumed to have an exponential distribution. Let $h(t)$ be the density function. We have

$$h(t) = \mu e^{-\mu t} \qquad (5\text{-}5)$$

where $\mu > 0$, which is the average service rate of a call (1/second). The mean holding time of a call is thus $E[t] = 1/\mu$.

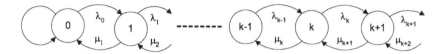

Figure 5-1. State Transition Diagram for a Birth–Death Process

5.1.2 Birth–Death Process

A Markov process is a simple stochastic process in which the distribution of future state depends only on the current state and not on the previous states. A Markov process with a discrete state space is referred to as a Markov chain. Let random variable $\{X_n\}$ denote the set of states. We have

$$\Pr\{X(t_{n+1}) = x_{n+1} \mid X(t_n) = x_n, \ldots, X(t_1) = x_1\}$$
$$= \Pr\{X(t_{n+1}) = x_{n+1} \mid X(t_n) = x_n\}$$
(5-6)

A birth–death process is a special case of a Markov process in which transitions are allowed only between adjacent states. Let S_k denote the state of a system in which the population (number of busy slots) is k. A transition from S_k to S_{k+1} with rate λ_k implies a birth in the population, i.e., an increase in the number of busy slots. A transition from S_k to S_{k-1} with rate μ_k implies a death in the population, i.e., a decrease in the number of busy slots. The state transition diagram of a birth–death process is depicted in Figure 5-1.

Let p_k be the probability that the system is in state S_k. The system reaches to equilibrium where the flow rate into state S_k equals the flow rate out of S_k Thus we have

$$\lambda_{k-1} p_{k-1} + \mu_{k+1} p_{k+1} = (\lambda_k + \mu_k) p_k, \ k \geq 0$$
(5-7)

where $\lambda_{-1} = 0$, $\mu_0 = 0$, and $p_{-1} = 0$.
Since

$$\sum_{k=0}^{\infty} p_k = 1$$
(5-8)

by solving the set of equations in (5-7), we have

$$p_k = p_0 \prod_{i=0}^{k-1} \lambda_i / \mu_{i+1} \tag{5-9}$$

and

$$p_0 = \left(1 + \sum_{k=1}^{\infty} \prod_{i=0}^{k-1} \lambda_i / \mu_{i+1} \right)^{-1} \tag{5-10}$$

Assuming that the arrival (birth) rate is a constant of λ and the departure (death) rate is a constant of μ, we have

$$p_k = p_0 \left(\lambda / \mu \right)^k \qquad p_0 = 1 - \lambda / \mu \tag{5-11}$$

5.1.3   Lost Call Cleared and Lost Call Held

There are two commonly used models for determining the state occupancy distribution and blocking probability in wireless systems, the "lost call cleared" (LCC) and "the lost call held" (LCH) models.

In the "lost call cleared" model, the system is assumed to have a finite number of channels (servers), and a new call will be lost if all the channels are busy at the call arrival time. Hence, the state transition diagram of Figure 5-1 terminates at state K (assuming total of K channels.) Assuming that the arrival rate and the departure rate of each state are

$$\begin{aligned} \lambda_k &= \lambda, \quad 0 \le k \le K - 1 \\ \mu_k &= k\mu, \quad 1 \le k \le K \end{aligned} \tag{5-12}$$

Applying equation   (5-12) to equations (5-9) and (5-10), we have

$$p_k = p_0 \frac{\left(\lambda / \mu \right)^k}{k!}, \qquad k = 1, 2, \ldots, K \tag{5-13}$$

$$p_0 = \left(\sum_{i=0}^{K} \frac{\left(\lambda / \mu \right)^i}{i!} \right)^{-1} \tag{5-14}$$

Combing (5-13) and (5-14), we obtain the occupancy distribution

$$p_k = \frac{\dfrac{(\lambda/\mu)^k}{k!}}{\displaystyle\sum_{i=0}^{K} \frac{(\lambda/\mu)^i}{i!}}, \qquad k = 0, 1, \ldots, K \tag{5-15}$$

The blocking probability is the probability that a new call arrives and finds that all the channels are busy, which is

$$p_B = p_K = \frac{\dfrac{(\lambda/\mu)^K}{K!}}{\displaystyle\sum_{i=0}^{K} \frac{(\lambda/\mu)^i}{i!}} \tag{5-16}$$

The expression in (5-16) is the well-known *Erlang B formula*, which represents the lost call or blocking probability in the LCC model. The computation of the Erlang B formula is well formed via a table.

In the "lost call held" (LCH) model, the system is assumed to have a finite number of channels (servers). A new call arrives and remains in the system unserved if all the channels are found to be busy. During the waiting period, if a channel becomes free, the waiting call seizes the channel and gets served. In this case, the number of states in the Markov chain is infinite.

Applying (5-12) to (5-9) and (5-10), we obtain the occupancy distribution

$$p_k = \frac{(\lambda/\mu)^k}{k!} e^{-\lambda/\mu}, \qquad k = 0, 1, 2, \ldots \tag{5-17}$$

The blocking probability is the probability that a new call arrives and finds all K channels are busy, which is the sum of the tail of a Poisson distribution.

$$p_B = \sum_{k=K}^{\infty} p_k = \sum_{k=K}^{\infty} \frac{(\lambda/\mu)^k}{k!} e^{-\lambda/\mu} \tag{5-18}$$

The LCH model has been used to represent the unslotted spread spectrum multiple access. It should be noted that for large K, the two models produce similar results.

5.2　Erlang Capacity for Circuit-Switched Services

For circuit-switched services, such as voice and circuit-switched data, a dedicated physical channel is assigned to a call for the entire call holding time. Therefore, traditional queuing analysis is commonly applied to evaluate the Erlang capacity of a wireless system. As noted previously, however, the capacity or number of users supported in a CDMA system depends on system resource, such as total bandwidth, data rate, interference limits, power limits, etc. The capacity analysis for reverse link and forward link with consideration of the various affecting factors is described in this section.

5.2.1　Capacity Analysis on Reverse Link

The CDMA system is strictly interference limited. On reverse link, many mobiles access to one base station. Each mobile transmits at a different power level with variable rate. For example, a voice call usually has different rates based on voice activity detection. The transmit power level is adjusted according to the voice activity. The received signal from each mobile shall meet a certain signal-to-noise ratio such that reliable information can be detected. On the other hand, the received signal shall be small such that the interference to other mobiles is kept at a minimum. To limit the interference and achieve the maximum capacity, power control is explored to control the received power from each mobile at a proper level.

Assuming that there are k_u active users in the current cell and k_{other} active users controlled by other cells, the total received power at the current cell is

$$\text{Total power} = \sum_{i=1}^{k_u} E_{bi} R_i + \sum_{j=1}^{k_{other}} E_{bj} R_j + N_0 W \tag{5-19}$$

where W is the total bandwidth (in Hz); N_0 is the background noise density (in watts/Hz); R_i (in bits/second) is the data rate of user i; E_{bi} (in watts/bit) is the received bit energy of user i.

The total received power is the sum of the desired user power, noise, and interference power. For one particular mobile (which is denoted by subscript

1), the total received noise plus interference power at the base station, denoted by I_0W, is

$$I_0W = \sum_{i=2}^{k_u} E_{bi} R_i + \sum_{j=1}^{k_{others}} E_{bj} R_j + N_0 W \qquad (5\text{-}20)$$

The first term in the right side of equation (5-20) represents the intracell interference that is introduced by other users in the same cell, while the second term represents the intercell interference that is introduced by users in all other cells.

The loading of the cell, Y, is defined as

$$Y = \frac{1}{W}\left(\sum_{i=1}^{k_u} \frac{E_{bi}}{I_0} R_i + \sum_{j=1}^{k_{other}} \frac{E_{bj}}{I_0} R_j\right) \qquad (5\text{-}21)$$

which is

$$Y = 1 - \frac{N_0}{\left(I_0 + E_{b1} R_1 / W\right)} \qquad (5\text{-}22)$$

The ratio of the overall interference plus noise power to background noise equals I_0W/N_0W, or equivalently I_0/N_0, which is

$$\frac{I_0}{N_0} = 1 + \frac{1}{W}\left(\sum_{i=2}^{k_u} \frac{E_{bi}}{N_0} R_i + \sum_{j=1}^{k_{other}} \frac{E_{bj}}{N_0} R_j\right) \qquad (5\text{-}23)$$

This ratio is also called the rise-over-thermal (ROT), which is an important metric to maintain system stability on reverse link. Because of the dynamic range limitation at the receiver as well as guarantee of system stability, it is desirable to limit the ROT within a certain range. Thus we have

$$\frac{I_0}{N_0} < \frac{1}{\eta}, \quad 0 < \eta < 1 \qquad (5\text{-}24)$$

Typically η is between 0.25 and 0.1, which corresponds to ROT of 6 dB to 10 dB. It can be adjustable based on the system requirement and implementation.

The system outage probability is defined as the probability that ROT exceeds a certain threshold. Thus we have

$$P_{out} = \Pr\left\{\frac{I_0}{N_0} > \frac{1}{\eta}\right\}$$

$$= \Pr\left\{\left(\sum_{i=2}^{k_u} E_{bi}R_i + \sum_{j=1}^{k_{other}} E_{bj}R_j\right) > I_0 W (1-\eta)\right\}$$

(5-25)

We further slightly increase the outage probability by including the desired signal variable in the first summation. Therefore the outage probability is

$$P_{out} = \Pr\left\{\left(\sum_{i=1}^{k_u} E_{bi}R_i + \sum_{j=1}^{k_{other}} E_{bj}R_j\right) > I_0 W (1-\eta)\right\}$$

(5-26)

In this expression, the number of active users k_u is a Poisson random variable with mean of $\alpha\lambda/\mu$ based on the LCH model [Vit93] where α is the activity factor, λ is the average call arrival rate and $1/\mu$ is the average call holding time. The objective is to estimate the Erlang capacity (λ/μ) that the system can provide under the system outage probability.

5.2.1.1 Erlang Capacity with Perfect Power Control for Single Type of Service

Users controlled by other cells introduce interference to the current cell. The other cell interference can be approximated by a factor f. Assuming uniform loading of every cell, the average other cell interference is equivalent to the received power from fk_u users. Because of soft handoff, the received signal from other cell users is always lower than the signal from the same cell users. Therefore, f is always below 1. For simplicity, the effect of other cell interference is modeled as an additional number of users in the current cell with a mean value of fk_u. Equation (5-26) becomes

$$P_{out} = \Pr\left\{ \sum_{i=1}^{(1+f)k_u} E_{bi}R_i > I_0W(1-\eta) \right\} \qquad (5\text{-}27)$$

Assuming that every mobile is perfectly power controlled by the base station, the received signal power is thus fixed at a certain level. For single type of service, the user data rate and the desired signal-to-interference ratio would be the same for every user. Let $R_i = R$ and $E_{bi} = E_b$ in equation (5-27). The outage probability is

$$
\begin{aligned}
P_{out} &= \Pr\left\{ k_u(1+f)E_bR > I_0W(1-\eta) \right\} \\
&= \Pr\left\{ k_u > \frac{(W/R)(1-\eta)}{(1+f)E_b/I_0} \right\} = \Pr\left\{ k_u > K_0 \right\}
\end{aligned}
\qquad (5\text{-}28)
$$

Based on the LCH model described in previous section, the outage probability in (5-28) is just the sum of the tail of the Poisson distribution, which is

$$P_{out} < e^{-\alpha\lambda/\mu} \sum_{k=\lfloor (1+f)K_0 \rfloor}^{\infty} \frac{(\alpha\lambda/\mu)^k}{k!} \qquad (5\text{-}29)$$

where

$$K_0 = \frac{(W/R)(1-\eta)}{E_b/I_0} \qquad (5\text{-}30)$$

5.2.1.2 Erlang Capacity with Imperfect Power Control

Power control intends to control the received signal from a mobile to a desired E_b/I_0 value. Based on analytical discussion and also experimental measurements, the received signal power varies according to a log-normal distribution. Therefore, the effect of imperfect power control can be taken into account by treating E_{bi} as a log-normal random variable. Let

$$E_{bi} = \varepsilon_i E_{bi0} \qquad (5\text{-}31)$$

where E_{bi0} is the desired bit energy, ε_i is a lognormal distributed random variable. Replacing the constant value of E_{bi} in equation (5-27) by a variable, we have

$$P_{out} = \Pr\left\{ \sum_{i=1}^{(1+f)k_u} \varepsilon_i > \frac{(W/R)(1-\eta)}{E_{b0}/I_0} \right\}$$ (5-32)

As noted earlier, $(1+f)k_u$ is a Poisson random variable. The mean of k_u is

$$\rho = \alpha \frac{\lambda}{\mu}$$ (5-33)

Let y_i denote the random variable of power control error in dB. We have

$$y_i = 10\log\varepsilon_i = \frac{1}{\beta}\ln\varepsilon_i, \quad \text{where } \beta = \ln 10/10$$ (5-34)

which is normally distributed, with zero mean and standard deviation σ_c. Therefore the mean and variance of the i.i.d. log-normal distributed random variable ε_i are

$$m_\varepsilon = e^{(\beta\sigma_c)^2/2}$$
$$\sigma_\varepsilon^2 = e^{2(\beta\sigma_c)^2} - m_\varepsilon^2$$ (5-35)

1. Voice Service Capacity

For voice services, the number of users in a cell is usually very high. Thus in equation (5-32), the left side of the inequality can be approximated as a Gaussian random variable. Let

$$x = \sum_{i=1}^{(1+f)k_u} \varepsilon_i$$ (5-36)

be a normally distributed random variable. The mean and variance of x is thus

$$m_x = (1+f)\rho m_\varepsilon$$
$$\sigma_x^2 = (1+f)\rho\sigma_\varepsilon^2$$ (5-37)

where m_ε and σ_ε are the mean and standard deviation of the *i.i.d.* log-normal random variable ε_i which are given in equation (5-35).

From (5-32), the outage probability can be easily derived as

$$P_{out} = Q\left(\frac{K_0 - (1+f)\rho m_\varepsilon}{\sqrt{(1+f)\rho}\sigma_\varepsilon}\right)$$

(5-38)

where $Q(\cdot)$ is the error function defined as

$$Q(x) = \frac{1}{\sqrt{2\pi}}\int_x^\infty e^{-x^2/2}dx$$

(5-39)

2. Circuit Switched Data Service Capacity

For circuit switched data services, the number of users is usually very small. Gaussian approximation of $\sum_{i=1}^{(1+f)k_u}\varepsilon_i$ will not be valid any longer. Since ε_i is log-normally distributed, the distribution of the sum of a finite number of log-normal random variables is well approximated by another log-normal distribution [Sch92]. Therefore, $\sum_{i=1}^{(1+f)k_u}\varepsilon_i \approx x$ is well modeled as a log-normal distributed random variable. Assuming the variance of ε_i is small, we can use Wilkonson's approach (which was summarized in [Sch92]) to calculate the mean and variance of x, which is

$$m_x = (1+f)\rho e^{m_\varepsilon + \sigma_\varepsilon^2/2}$$
$$\sigma_x^2 = (1+f)\rho e^{2m_\varepsilon + 2\sigma_\varepsilon^2}$$

(5-40)

By further derivation, the outage probability for circuit switched data is

$$P_{out} = Q\left(\frac{\ln(K_0) - m_{\ln x}}{\sigma_{\ln x}}\right)$$

(5-41)

where $x = \displaystyle\sum_{i=1}^{(1+f)k_u} \varepsilon_i$, and

$$\sigma_{\ln x}^2 = \ln\left(\frac{\sigma_x^2}{m_x^2} + 1\right)$$

$$m_{\ln x} = \ln(m_x) - \frac{1}{2}\sigma_{\ln x}^2$$

(5-42)

5.2.1.3 Erlang Capacity with Imperfect Power Control and Intercell Interference Modeling

For single type of service with imperfect power control, we can rewrite equation (5-26) as

$$P_{out} = \Pr\left\{\left(\sum_{i=1}^{k_u}\varepsilon_i + \sum_{j=1}^{k_{other}}\varepsilon_j\right) > \frac{(W/R)(1-\eta)}{E_{b0}/I_0}\right\}$$

(5-43)

As discussed in the previous section, because of the nature of soft handoff, users in other cells can not arrive at the given cell base station with a higher power than the same cell users, i.e., $E[\varepsilon_j] < 1$. Therefore the intercell interference term in equation (5-43) can be modeled as $fE[k_u]$, which is equivalently the added average number of active users in the given cell because of interference introduced by all othercell users. As discussed in precious section, f is defined as the intercell to intracell interference ratio and is less than 1. The outage probability is thus

$$P_{out} = \Pr\left\{\left(\sum_{i=1}^{k_u}\varepsilon_i + fE(k_u)\right) > K_0\right\}$$

(5-44)

and

$$K_0 = \frac{(W/R)(1-\eta)}{E_{b0}/I_0}$$

(5-45)

The effect of the intercell interference is modeled by a fixed intercell to intracell interference ratio f in previous sections. Because of soft handoff, this ratio is always less than 1. Its value depends on the propagation path loss model and the standard deviation of the log-normal shadow fading.

Using a fixed value of f simplifies the outage probability calculation. However it does not capture the dynamics of the intercell interference in reality and thus gives an inaccurate estimation.

The analysis can be improved by treating the intercell interference ratio f as a random variable. By examining the path loss and shadow fading distribution, it is proved that f can be characterized by a Gaussian or log-normal distributed random variable, depending on the loading of other cells. When the cell loading is high, f can be approximated as a Gaussian distributed random variable. When the cell loading is low, f is better approximated as a log-normal distributed random variable. Simulations were performed to verify the accuracy of the modeling of intercell interference.

1. Voice Service Capacity

We re-evaluate the capacity calculation considering the intercell interference modeling. Because of large number of users, we use Gaussian approximation to simplify the calculation of the outage probability. Let

$$x = \sum_{i=1}^{k_u} \varepsilon_i + fE[k_u] \tag{5-46}$$

be a Gaussian random variable with mean and variance as

$$
\begin{aligned}
m_x &= \rho m_\varepsilon + \rho m_f \\
\sigma_x^2 &= \rho \sigma_\varepsilon^2 + \rho^2 \sigma_f^2
\end{aligned}
\tag{5-47}
$$

where m_f and σ_f are the mean and standard deviation of f; m_ε and σ_ε are defined in (5-35). The outage probability can be easily derived as

$$P_{out} = Q\left(\frac{K_0 - (\rho m_\varepsilon + \rho m_f)}{\sqrt{\rho \sigma_\varepsilon^2 + \rho^2 \sigma_f^2}} \right) \tag{5-48}$$

2. Circuit Switched Data Service Capacity

Following the same analysis in section 5.2.1.2, the sum of the received signal is approximated as a log-normal distributed random variable x. The inter-cell to intra-cell interference ratio f is also approximated as a log-normal distributed random variable. Again, the summation of $(x + \rho f) \approx z$ is modeled as a log-normal distributed random variable and its mean and variance can be calculated. By further derivation, the outage probability for circuit switched data is

$$P_{out} = Q\left(\frac{\ln(K_0) - m_{\ln z}}{\sigma_{\ln z}} \right) \tag{5-49}$$

where $z = \sum_{i=1}^{k_u} \varepsilon_i + \rho f$ and

$$\sigma_{\ln z}^2 = \ln\left(\frac{\sigma_z^2}{m_z^2} + 1 \right)$$

$$m_{\ln z} = \ln(m_z) - \frac{1}{2}\sigma_{\ln z}^2 \tag{5-50}$$

with

$$m_x = \rho e^{m_\varepsilon + \sigma_\varepsilon^2/2} + \rho m_f$$

$$\sigma_x^2 = \rho e^{2m_\varepsilon + 2\sigma_\varepsilon^2} + \rho^2 \sigma_f^2 \tag{5-51}$$

5.2.1.4 Capacity Model with Imperfect Power Control for Mixed Types of Services

For mixed type of services, the required signal-to-interference ratio and data rate for each service is different. We consider the case of two types of services, for example, voice and data services with rate R_1 and R_2, and bit energy to interference ratio $(E_{b0}/I_0)_1$ and $(E_{b0}/I_0)_2$. The case of more than two service types can be extended easily. We can rewrite the outage probability in equation (5-26) as

$$P_{out} = Pr\left\{ C_1\left(\sum_{i=1}^{k_1} \varepsilon_i + \rho_1 f_1 \right) + C_2\left(\sum_{j=1}^{k_2} \varepsilon_j + \rho_2 f_2 \right) > (1 - \eta) \right\} \tag{5-52}$$

where

$$C_{1,2} = \frac{\left(E_{b0}/I_0\right)_{1,2}}{W/R_{1,2}}$$ (5-53)

$$\rho_{1,2} = \alpha_{1,2}\frac{\lambda_{1,2}}{\mu_{1,2}}$$ (5-54)

In this expression, the variables related with voice users are modeled as Gaussian distributed random variables, while those related with data users are modeled as log-normal distributed random variables. Following the same procedure of deriving the probability density function of multiple random variables, we can calculate the outage probability numerically.

5.2.2 Capacity Analysis on Forward Link

Compared to reverse link, the analysis on the forward link is much more challenging owing to the very different implementation on the forward link. For forward link, the access is one-to-many instead of many-to-one. The interference each mobile has is received from a few concentrated large sources (base stations) rather than many distributed small mobiles. Base station transmission power is shared by many users and is allocated according to each user's relative need. The forward link operation is rather a procedure as forward power allocation so that many users can be served and a cell coverage can be provided as well. Because of the difficulties in analyzing forward link capacity, most of the performance evaluation was obtained through simulations. Following the development of [Zha2_01], we describe a framework to analyze the system capacity on the forward link. Particularly, we present a novel approach to model the interference distribution and derive simple expressions to calculate the system capacity quickly and effectively. In the analysis, we also include soft handoff and evaluate its impact on the system capacity. Furthermore, we apply different approximation methods to estimate capacity based on voice and/or data traffic loading.

5.2.2.1 Analytical Model for Forward Links

As described earlier, the forward link performance is very different from that of the reverse link in CDMA systems. The total transmission power is the system resource shared by many users. Assume that user i is controlled by base station 1. It receives interference from other base stations. Ideally, signals transmitted to mobiles in the same cell are orthogonal so that they will not interfere with each other. However because of multipath propagation, the received signals are no longer perfectly orthogonal and thus cause intracell interference. We use an orthogonality factor to represent this effect. For user i, the received bit energy to interference plus noise density ratio is

$$\left(\frac{E_b}{I_0}\right)_i = \frac{\phi_i S_{R_{1i}} / R_i}{\left(\sum_{j=2}^{J} S_{R_{ji}} + (1 - f_{orth}) S_{R_{1i}} + N_0 W\right) / W} \tag{5-55}$$

where ϕ_i is the fraction of the total transmitted power allocated to user i; J is the number of base stations; S_{Rji} is the total power received by user i from base station j; f_{orth} is the orthogonality factor, and $0 \leq f_{orth} \leq 1$; W is the total bandwidth; N_0 is the background noise density; R_i is the data rate of user i. We assume that the user always communicates with the base stations (one or multiple base stations in soft handoff situation) from which it receives the strongest signal. In other words, $S_{R1i} > S_{Rji}$, for all $j \neq 1$.

In practical systems, a fraction of the total transmission power is devoted to the pilot signal and other common control channels destined to all users. We assume this overhead is $(1-\beta)$. Therefore the remaining fraction β of the total power is then allocated to all users controlled by the base station in the cell or sector. The system will be in an outage situation if the total allocated transmission power exceeds the total available power at the base station. We define the system outage probability as

$$P_{out} = \Pr\left\{\sum_{i=1}^{k_u} \phi_i > \beta\right\} \tag{5-56}$$

where k_u is the number of active users in the current cell or sector. By substituting ϕ_i from equation (5-55) to (5-56), we obtain the outage probability as

$$P_{out} = \Pr\left\{\sum_{i=1}^{k_u} \frac{(E_b / I_0)_i}{W / R_i}\left(\sum_{j=2}^{J} \frac{S_{R_{ji}}}{S_{R_{1i}}} + (1 - f_{orth}) + \frac{N_0 W}{S_{R_{1i}}}\right) > \beta\right\} \tag{5-57}$$

As discussed in Section 5.2.1, the number of active users k_u is a Poisson random variable with mean of $\alpha\lambda/\mu$ where α is the activity factor, λ is the average call arrival rate, and $1/\mu$ is the average call holding time. The objective is to estimate the Erlang capacity (λ/μ) that the system can provide below a system outage probability.

5.2.2.2 Capacity Analysis with No Soft Handoff for Single Type of Service

For single type of service, the user data rate and the desired signal-to-interference ratio are the same for every user. We assume perfect power control on the forward link. In general, the background noise is negligible compared to the total power received from all base stations. We can thus drop N_0W in equation (5-57). Let

$$y_i = \sum_{j=2}^{J} S_{R_{ji}} / S_{R_{1i}} \tag{5-58}$$

The outage probability is thus

$$P_{out} = \Pr\left\{\sum_{i=1}^{k_u}(y_i + 1 - f_{orth}) > \beta K_s\right\} \tag{5-59}$$

where

$$K_s = \frac{W / R_s}{(E_b / I_0)_s} \tag{5-60}$$

and the subscript s indicates the service type.

5.2.2.3 Modeling of Interference from Other Base Station

In (5-58), y_i depends on user i's location and log-normal shadow fading variables. This distribution of y_i is not tractable analytically. In previous work, the distribution of y_i was obtained through simulations and no analytical formula was provided. In [Zha2_01], an approximation technique was developed to model this interference distribution so that the outage probability can be calculated effectively.

We assume each base station transmits at the same power level. The total power S_{Rji} received by user i from the jth base station is proportional to the inverse of the path loss plus shadow fading which is log-normally distributed. Thus S_{Rji} is a log-normal random variable. The ratio S_{Rji}/S_{R1i} is also log-normally distributed. The variable y_i defined in (5-58) is the sum of a finite number of log-normal random variables. Its distribution is well approximated by another log-normal distribution [Sch92]. Since the variance of S_{Rji} is large (usually larger than 6 *dB*), we follow the technique presented in [Sch92] to estimate the mean and variance of the sum of log-normal components, y_i. In summary, the procedure is to compute the mean and variance using the derived analytical formulae for the sum of two components, then to iterate until the required moments are obtained.[3]

We should note that the distribution of y_i depends on the position of the *i*th user. For the user at different locations, the relative distances from other base stations will be different and thus the estimated mean and variance of y_i will not be the same. We divide the cell into equally spaced points and calculate the mean and variance of y_i for each location.

1. Voice Service Capacity

For voice service, the number of users k_u is usually very large in a CDMA system. In equation (5-59), let

$$z = \sum_{i=1}^{k_u} \left(y_i + 1 - f_{orth} \right) \qquad (5\text{-}61)$$

which can be approximated as a Gaussian random variable. Since y_i depends on the user's position, we approximate the estimation by taking the average mean and variance of y_i at all the locations in the cell (or sector), assuming that users are uniformly distributed in the cell. Let $\rho = E[K_u] = \alpha \dfrac{\lambda}{\mu}$ where α is the voice activity factor and λ/μ is the loading of voice calls. The mean and variance of z can thus be approximated as

$$m_z = \rho \left(m_y + 1 - f_{orth} \right)$$
$$\sigma_z^2 = \rho \sigma_y^2 + \rho \left(m_y + 1 - f_{orth} \right)^2 \qquad (5\text{-}62)$$

From (5-59) and (5-62), the outage probability can be derived as

---

[3] Interested readers can refer [Sch92] for the detailed formulae of the calculation, which are equations (22)–(24), (30) – (32) and Tables 1a, 1b.

$$P_{out} = Q\left(\frac{\beta K_v - \rho\left(m_y + 1 - f_{orth}\right)}{\sqrt{\rho\sigma_y^2 + \rho\left(m_y + 1 - f_{orth}\right)^2}} \right) \tag{5-63}$$

where $Q(\cdot)$ is the error function; m_y and σ_y are the mean and standard deviation of y_i by averaging over all positions that are given in the following section.

2. Circuit Switched Data Service Capacity

For circuit switched data services, the number of users is usually very small. Since the variable z is the sum of log-normal random variables, we can model z as another log-normally distributed random variable. The mean and variance of z are difficult to tract analytically because of the presence of multiple random variables, i.e., Poisson random variable k_u and location-dependent log-normal random variable y_i. For simplicity, we treat each y_i as an i.i.d. random variable with the same average mean and variance. The mean and variance of z can thus be estimated via equation (5-62).

By further derivation, the outage probability for circuit switched data is

$$P_{out} = Q\left(\frac{\ln\left(\beta K_d\right) - m_{\ln(z)}}{\sigma_{\ln(z)}} \right) \tag{5-64}$$

where

$$\sigma_{\ln(z)}^2 = \ln\left(\frac{\sigma_z^2}{m_z^2} + 1 \right)$$

$$m_{\ln(z)} = \ln\left(m_z\right) - \frac{1}{2}\sigma_{\ln(z)}^2 \tag{5-65}$$

and m_z and σ_z are given by equation (5-62).

5.2.2.4 Capacity Model with No Soft Handoff for Mixed Types of Services

For mixed type of services, the required signal-to-interference ratio and data rate for each service is different. We consider the case of two types of services, for example, voice and data services with rate R_v and R_d, and bit energy to interference ratio $(E_b/I_0)_v$ and $(E_b/I_0)_d$. The case of more than two service types can be extended easily. We can rewrite the outage probability in equation (5-59) as

$$P_{out} = \Pr\left\{\frac{1}{K_v}\sum_{i=1}^{k_v}(y_i + 1 - f_{orth}) + \frac{1}{K_d}\sum_{j=1}^{k_d}(y_j + 1 - f_{orth}) > \beta\right\} \quad (5\text{-}66)$$

where

$$K_{v,d} = \frac{W / R_{v,d}}{\left(E_b / I_0\right)_{v,d}} \quad (5\text{-}67)$$

Similarly, the variable related with voice users can be approximated as a Gaussian distributed random variable Z_v, while the one related with data users can be approximated as a log-normal distributed random variables Z_d. The outage probability is thus

$$P_{out} = \Pr\left\{\frac{1}{K_v}Z_v + \frac{1}{K_d}Z_d > \beta\right\} \quad (5\text{-}68)$$

Following the procedure of deriving the probability density function of multiple random variables, we can calculate the outage probability numerically.

5.2.2.5 Capacity Model with Soft Handoff

Soft handoff has very different impacts on reverse link and forward link. Soft handoff improves reverse link performance considerably with virtually no interference drawbacks. It provides large diversity gain and is almost free for the mobile user since the second base station receives signals from the mobile anyway. Soft handoff on the forward links is a different situation. The second base station must also transmit the same signals to the user as the first base station. The transmitted power is therefore no longer available to other users in the second cell or sector. As a result, soft handoff actually impairs the system capacity on the forward links. Of course, the advantage of soft handoff is still evident on the forward links because of the diversity gain. It provides a smooth transition for hard handoff, especially that ideal instantaneous hard handoff is almost never met under the rapidly varying environment.

Soft handoff on the forward links makes the power allocation problem even more complicated. For simplicity, we assume that a fraction $g < 1$ of all the users is in soft handoff and maximum two base stations will get involved (the number of legs is 2). For each user in soft handoff, we assume that both base stations allocate the same power fraction to that user. Effectively, the number of users in each cell is increased by a fraction g because of soft handoff. Hence, the cell loading in outage probability derivation of equations (5-57) and (5-68) will increase from λ/μ to $(\lambda/\mu)(1+g)$. The system capacity with soft handoff can then be determined.

5.3 Capacity for Packet Switched Services

Circuit switching is a type of communication method where a dedicated channel (or circuit) is established for the whole duration of the transmission. In contrast to circuit switching, packet switching is another common communication method where messages are divided into packets and each packet is sent individually. Circuit switching is ideal for services that require data to be transmitted in real time. Packet switching is more efficient for services if some amount of delay is acceptable.

The current 3G wireless systems support both circuit switched and packet switched services. Packet data users share the air interface channels and make efficient use of the resources. If the data user is idle and not transmitting, the channel is released and assigned to another user for transmission. The base station schedules the transmission of data users trying to maximize the system capacity and also achieve fairness for all users at the same time.

The capacity analysis for packet data services is more challenging than that for circuit-switched services because of the dynamic resource allocation for data users. Traffic process and characteristics for packet data applications are more diverse and complicated than traditional voice traffic. The technique of hybrid ARQ, dynamic rate control and scheduling discipline makes the system difficult to analyze. Simple and approximate analysis has been used to evaluate the reverse link capacity for high-speed data services. The analysis is basically the same as that for circuit switched services and thus can provide only an approximate capacity (or throughput) estimation. Because of those challenges, simulations are often used to assess the capacity for packet switched services. In the following, we describe more details on the simulation methodologies for capacity evaluation.

5.4   Simulation Methodologies for Capacity Evaluation

System or cell level simulation is widely used to evaluate the air interface capacity. Link level simulation is performed separately and generates the performance characteristics of individual links such as the required E_b/I_0 for different services. These results are generated a priori and used as the inputs to the system level simulation. There are usually two types of system level simulations: static simulation and dynamic simulation. In static simulation, mobiles are randomly placed in multiple cells in one snapshot. Path loss and fading statistics are collected for each mobile at each run. The simulation runs enough number of snapshots so that the outage probability can be calculated. There is no call arrival and departure modeled in the static simulation. Static simulation is often used to evaluate the capacity for voice and circuit-switched data services. For the circuit switched services, a dedicated physical channel is assigned for each mobile. The static snapshot approach is accurate enough to estimate the outage probability. In dynamic simulation, mobiles are randomly placed in multiple cells initially and generate packet arrival and departure continuously. The simulation includes path loss and fading and evolves in time with discrete steps. Dynamic simulation is often used to evaluate the packet data system performance. It can accurately model the feedback loop, random packet arrivals and departures, user scheduling, and signal transmission and reception latency.

5.4.1   System Level Simulation Assumptions for Forward Link

The commonly used parameters and assumptions for forward link in system level simulation are shown in Table 5-1. The values in the table are recommended by the 3GPP2 standards [3GPP2].

The services evaluated in system level simulations are described in the traffic models. Common traffic models include voice, Web browsing, FTP, streaming video and audio, etc. The details of the traffic model are described in Chapter 3.

Table 5-1 Parameters for Forward Link System Level Simulations [3GPP2]

Parameter	Value	Comments
Number of cells (3 sectored)	19	2 rings, 3-sector system, 57 sectors.
Antenna horizontal pattern	70 deg (-3 dB) with 20 dB front-to-back ratio	
Antenna orientation	The 0 degree horizontal azimuth is east (main lobe).	No loss is assumed on the vertical azimuth.
Propagation model (BTS Ant Ht = 32 m, MS = 1.5 m)	$28.6 + 35\log10(d)$ dB, d in meters	Modified Hata Urban Prop. Model @1.9GHz (COST 231). Minimum of 35 m separation between MS and BS[4]
Log-normal shadowing	Standard deviation = 8.9 dB	Independently generate log-normal per mobile
Base station correlation	0.5	
Overhead channel forward link power usage	Pilot, Paging and Sync overhead: 20%.	Any additional overhead needed to support other control channels (dedicated or common) must be specified and justified.
Mobile noise figure	10.0 dB	
Thermal noise density	−174 dBm/Hz	
Carrier frequency	2 GHz	
Base station antenna gain with cable loss	15 dB	17 dB base station antenna gain; 2 dB cable loss

---

[4] If a mobile is dropped within 35 m of a base station, it shall be redropped until it is outside the 35-m circle.

Table 5-1 Cont.

Parameter	Value	Comments
Mobile antenna gain	−1 dBi	
Other losses	10 dB	
Fast fading model	Based on speed	Jakes or Rician
Active set parameters		Secondary pilots within 6 dB of the strongest pilot and above minimum E_c/I_o threshold (−16dB). The active set is fixed for the drop. The maximum active set size is three.
Forward link power control (If used on dedicated channel)	Power control loop delay: two PCGs[5]	Update rate: Up to 800-Hz PC BER: 4%
Base station maximum PA power	20 Watts	
Site to site distance	2.5 km	
Maximum C/I achievable, where C is the instantaneous total received signal from the serving base station(s) (usually also referred to as $rx\_I_{or}(t)$, or $\hat{I}_{or}(t)$), and I is the instantaneous total interference level (usually also referred to as $N_t(t)$).	13 dB and 17.8 dB	13 dB for typical current subscriber designs for IS-95 and CDMA2000 1x systems; 17.8 dB for improved subscriber designs for 1xEV-DV systems.

[5] One PCG delay in link level modeling (measured from the time that the SIR is sampled to the time that the BS changes TX power level.)

5.4.2 System Level Simulation Assumptions for Reverse Link

The parameters and assumptions used in reverse link system level simulation are listed Table 5-2. The values are recommended by the 3GPP2 standards [3GPP2].

Table 5-2 Parameters for Reverse Link System Level Simulations [3GPP2]

Parameter	Value	Comments
Number of 3-sector cells	19	2 rings, 3-sector system, 57 sectors total, cells are on a "wrap-around" model.
Antenna horizontal pattern	70 degree (–3 dB) with 20 dB front-to-back ratio	
Antenna orientation	0 degree horizontal azimuth is East (main lobe)	No loss is assumed on the vertical azimuth.
Propagation model (BTS Ant Ht = 32 m, MS = 1.5 m)	$28.6 + 35\log10(d)$ dB, d in meters	Modified Hata Urban Prop. Model @1.9GHz (COST 231). Minimum of 35 m separation between mobile and base station.[6]
Log-normal shadowing	Standard deviation = 8.9 dB for both FL and RL	Independently generate lognormal per mobile-sector pair.
Maximum RL total path loss	146 dB	This term includes the mobile and base station antenna gains, cable and connector losses, other losses, and shadowing, but not fading.
Base station shadowing correlation	0.5	

[6] If a mobile is dropped within 35 m of a base station, it shall be redropped until it is outside the 35-m circle.

Table 5-2 Cont.

Parameter	Value	Comments
Overhead channel reverse link power usage		Existing IS-2000 traffic channel to pilot channel power ratio defined as in IS-2000. Any additional overhead needed to support other control channels (dedicated or common) for the forward link or the reverse link must be specified and justified
Base noise figure	5.0 dB	
Thermal noise density	–174 dBm/Hz	
Carrier frequency	2 GHz	
Base station antenna gain with cable loss	15 dB	17 dB BS antenna gain; 2 dB cable loss
Mobile antenna gain	–1 dBi	
Other losses	10 dB	Applicable to all fading models
Fast fading model	Based on speed	The fading processes on the paths from a given mobile to the two base station antennas are mutually independent.
Active set membership		Up to 3 members are in the active set if the pilot E_c/I_o is larger than T_ADD = –18 dB (= 9 dB below the FL pilot E_c/I_{or}) based on the FL evaluation methodology
Delay spread model		ITU pedestrian A for 1 finger, vehicular A for 2 fingers, pedestrian B for 3 fingers

Table 5-2 Cont.

Parameter	Value	Comments
Active set change		System specific. Proponents need to declare the scheme and the associated signaling delay and reliability.
Reverse link power control	Closed-loop power control delay: two PCGs[7]	Update rate: dependent on proposal. Power control feedback: BER = 4% for a BS-MS pair. Different values shall be specified and justified Ec/Nt measurement error at the BS: additive in dB, log normal, zero-mean random variable with a 2 dB standard deviation.
MS PA size	200 mW	
Site to site distance	2.5 km	
Rise over Thermal (Reverse received power normalized by thermal noise level)	7 dB	Histogram of this parameter with a 1.25-ms time resolution shall be provided with the mean rise-over-thermal. The percentage of time the rise over thermal above the 7 dB target shall not exceed 1%. Rise over thermal for the default two receiving antenna mode is $\frac{1}{2}[(I_{o1}+N_o)/N_o + (I_{o2} + N_o)/N_o]$, where the total received signal power at antenna i is defined as I_{oi}, $i = 1, 2$.

---

[7] The mobile transmit power changes in PCG $i+2$ in response to measurement made in PCG i. One PCG delay for link level modeling (measured from the last chip that the reverse pilot is measured to the time that the mobile changes TX power level).

The traffic models for different services on the reverse link are described in Chapter 3.

5.4.3 Performance Criteria and Output Metrics

For circuit switched voice and data services, the performance criterion is that the system shall meet a certain outage probability. For example, 5% outage probability is often used as the system requirement. For packet switched data services, the performance criteria have more metrics other than the outage probability.

- Fairness criterion

Because maximum system capacity may be obtained by providing low throughput to some users, it is important that all mobile stations be provided with a minimal level of throughput. This is called fairness. The fairness is evaluated by determining the normalized cumulative distribution function (CDF) of the user throughput, which meets a predetermined function under some test conditions.

For example, in 3GPP2 standards, it is required that the CDF curve shall lie to the right of the curve given by the three points in Table 5-3.

Table 5-3 CDF Fairness Criteria [3GPP2]

Normalized Throughput w.r.t average user throughput	CDF
0.1	0.1
0.2	0.2
0.5	0.5

- Delay criterion

The delay criteria shall be satisfied by all packet data users or packet data users of a particular application. For example, no more than 2% of the users shall get less than 9.6 Kbps throughput. As another example, no more than 2% of the streaming video users shall get less than 16 Kbps throughput.

The commonly used output metrics are:
- Aggregate data throughput per sector

The data throughput of a sector is defined as the number of information bits per second that a sector can deliver and that are received successfully by all data users it serves.

- Average data throughput per user

The throughput of a user is defined as the ratio of the number of information bits that the user successfully receives during a simulation run and the simulation time.

- Average packet delay per sector

The average packet delay per sector is defined as the ratio of the accumulated delay for all packets it delivers to all users and the total number of packets it delivers. The delay for an individual packet is defined as the time between when the packet enters the queue at the transmitter and the time when the packet is received successfully by the mobile station.

- Average packet delay per user

The average packet delay per user is defined as the ratio of the accumulated delay for all packets it delivers to the user and the total number of packets it delivers. The delay for an individual packet is defined as the time between when the packet enters the queue at transmitter and the time when the packet is received successfully by the mobile station.

- Distribution of data throughput per user

The throughput of a user is defined as the ratio of the number of information bits that the user successfully receives during a simulation run and the simulation time.

- Distribution of average packet delay per user

The average packet delay is defined as the ratio of the accumulated delay for all packets for the user and the total number of packets for the user.

5.5 Comparison of Analytical Models with Simulations

To verify the analytical models developed in Section 5.2, simulations were carried out to estimate the capacity and compare with the analytical results. Simulation models for reverse link and forward link are built separately. We evaluate the capacity based on the parameters and assumptions for the UMTS systems. The results can be extended to other CDMA systems as well.

5.5.1   Comparison of Analytical and Simulation Results on Reverse Link

A static system level simulation was built for capacity evaluation. The cell and sector layout follows the assumptions described in Section 5.4. Each cell is a hexagon and has three sectors. We assume that mobile users are uniformly distributed among the 19 cells. The path loss and shadow fading model, and the link level simulation parameters were obtained from the UMTS standard documents [25.201][25.211]–[25.215]. On reverse link, each user communicated with the strongest base station and was power controlled by the serving base station. The received signal-to-noise ratio in dB at the base station was modeled as a log-normal random variable to reflect the imperfect power control effect. User activity was also simulated. A user would be in soft handoff if its received signal power from different base stations fell in the soft handoff window. Selective combining was performed when a user was in soft handoff.

The simulation was static where a "snapshot" approach was used. No mobility model was considered for this capacity estimation. For each trial, we generated a Poisson random number as the number of mobiles in the system. Although the number of mobiles changed every time, the mean value of the Poisson random number generator was fixed which represents the carried load (λ/μ in terms of Erlang) based on the LCH model. We tracked the intercell interference from mobiles in other sectors/cells and the intracell interference from mobiles served by the same sector. The statistics were collected from the center cell only. For each snapshot, an outage event happened if the total received interference at the given base station exceeded a certain threshold.[8] For each loading (in terms of Erlang), we ran simulations more than 10,000 times and calculated the outage probability. For mixed types of services, we generated two types of traffic and collected their statistics accordingly.

The simulation parameters are summarized in Table 5-4.

The capacity results for single type of service based on 5% system outage probability are stated in Table 5-5. It should be noted that the capacity results highly depend on the link level results and other assumptions and parameters. The table here is for reference only.

---

[8] Here we did not consider the mobile transmission power limit. We emphasized the system outage only.

Table 5-4 Reverse Link Simulation Parameters [Zha1_01]

Environment	Vehicular A
Cell radius	2 km
Sectors/cell	3
Chip rate (chips/sec)	3.84 M
Voice activity factor	0.5
Data activity factor	1
Number of carriers	1
Log-normal shadow fading standard deviation	8 dB
Soft handoff window	6 dB
Standard deviation of power control error	1.5 dB
Thermal noise density	−174 dBm/Hz
η	0.10

Table 5-5 Reverse Link Capacity for Circuit Switched Services [Zha1_01]

Service (bps)	E_b/I_0 [dB]	Block error rate	Capacity (Erlang/carrier/sector)
Voice 8k	5.4	1%	125
LCD[9]144k	3.1	10%	4
LCD64k	3.8	10%	9.7

Figure 5-2 shows the comparison of our analysis and simulation results for voice traffic, with perfect power control and imperfect power control. We used Gaussian approximation to model the imperfect power control effect. We see that analytical results match very well with simulation.

Figure 5-3 shows the effect of intercell interference modeling for the perfect power control case. In our analytical model, for voice traffic, the intercell to intracell interference ratio f was approximated by a Gaussian random variable with mean of 0.5544 and standard deviation of 0.0601, which were obtained from calculation and verified by simulations. Using a fixed value (which was 0.5544) to represent intercell interference gives optimistic capacity results. Instead, modeling inter-cell interference as a random variable shows very good results as compared with simulations.

---

[9] LCD stands for Low Constrained Delay service.

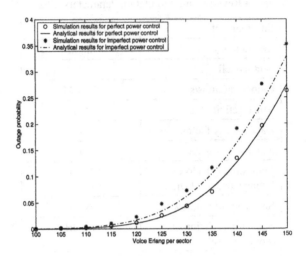

Figure 5-2 Comparison of Analysis and Simulation for voice traffic ©2001 IEEE
[Zha1_01].

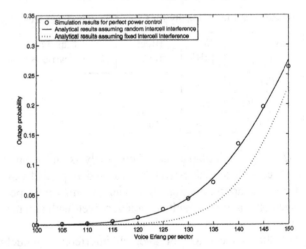

Figure 5-3 Comparison of Different Intercell Interference Modeling Approaches ©2001 IEEE
[Zha1_01]

We further compare our analytical model with simulations for data traffic, as shown in Figure 5-4. We compare different approximations for the imperfect power control effect. With log-normal approximation, the analysis shows a very good match with simulations. On the contrary, the Gaussian approximation does not work well and generates optimistic capacity results.

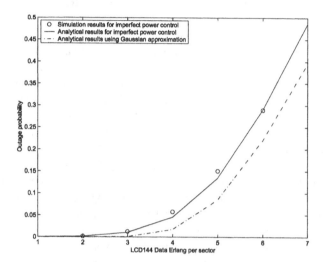

Figure 5-4 Comparison of Analysis and Simulation for Data Traffic ©2001 IEEE [Zha1_01]

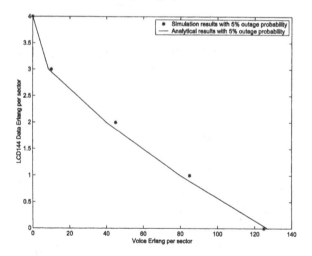

Figure 5-5 Comparison of Analysis and Simulation for Mixed Voice and Data Traffic ©2001 IEEE [Zha1_01]

Figure 5-5 shows the capacity for mixed voice and data services with 5% system outage probability. We see that the relationship between voice and data capacity is not linear. Again, our analysis matches well with simulations.

5.5.2   Comparison of Analytical and Simulation Results on Forward Link

A static system level simulation was built for forward link based on the UMTS standards specifications. On forward link, each user communicated with the strongest base station and was perfectly power controlled by the serving base station. A user would be in soft handoff if its received signal power from different base stations fell in the soft handoff window. Maximum ratio combining was performed when a user was in soft handoff. We assume that each base station was transmitting at the full power. We tracked the interference level (including both intercell and intracell interference) each active mobile received and allocated power according to its need. For each snapshot, an outage event happened if the total required transmission power exceeded the total available power.[10] For each loading (in terms of Erlang), we ran simulations more than 10,000 times and calculated the outage probability. For mixed types of services, we generated two types of traffic and collected their statistics accordingly.

Table 5-6 Forward Link Simulation Parameters [Zha2_01]

Environment	Vehicular A
Cell radius	2 km
Sectors/cell	3
Chip rate (chips/second)	3.84 M
Voice activity factor	0.5
Data activity factor	1
Number of carriers	1
Log-normal shadow fading Standard deviation	8 dB
Soft handoff window	6 dB
Number of soft handoff legs	2
Orthogonality factor	0.4
Maximum base station transmission power	20 Watt
Thermal noise density	−174 dBm/Hz
Overhead to pilot and other common control channels $(1-\beta)$	15%

The simulation parameters are summarized in Table 5-6.

---

[10] Here we emphasized on the system outage only. We did not consider the probability of mobiles in outage because of coverage.

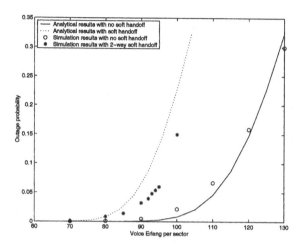

Figure 5-6 Comparison of Analysis and Simulation for Voice Traffic ©2001 IEEE
[Zha2_01]

The capacity results for single type of service based on 5% system outage probability are stated in Table 5-7. It should be noted that the capacity results highly depend on the link level results and other assumptions and parameters. The table here is for reference only.

Table 5-7 Forward Link Capacity for Voice and Data [Zha2_01]

Service (bps)	E_b/I_0 [dB]	Block error rate	Capacity (Erlang/carrier/sector)
Voice 8k	7.2	1%	93.5
LCD144k	2.5	10%	4.5
LCD64k	3.7	10%	9.8

Figure 5-6 shows the comparison of our analysis and simulation results for voice traffic, with and without soft handoff. The average mean and standard derivation of y_i over 375 points from numerical calculations were $m_y = 0.5767$ and $\sigma_y = 0.6764$. As shown in the figure, soft handoff actually hurts system capacity on the forward link. We see that analytical results match very well with simulation for no soft handoff case. With soft handoff, there is some discrepancy between analysis and simulation due to the simplified soft handoff modeling. The analytical results mainly depend on the percentage of mobiles in soft handoff. In our example we assume $g = 25\%$.

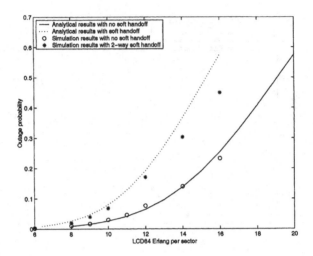

Figure 5-7. Comparison of Analysis and Simulation for Data Traffic ©2001 IEEE [Zha2_01]

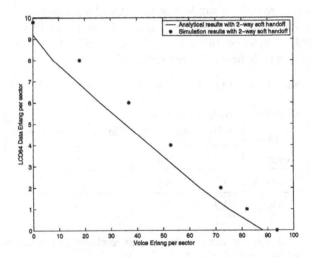

Figure 5-8. Comparison of Analysis and Simulation for Mixed Voice and Data Traffic ©2001 IEEE [Zha2_01]

Next we compare our analysis with simulations for data traffic, as shown in Figure . We use the log-normal approximation to model the variable Z and calculate the outage probability according to (5-64). The same values of m_y and σ_y were used. Again, the analysis shows a very good match with simulations for the no soft handoff case. With soft handoff, the analytical results have a reasonable good match with simulations.

Figure 5- shows the capacity for mixed voice and data services with 5% system outage probability with soft handoff. Our analysis has good agreement with simulations.

The analytical results have a very good match to the simulations. This demonstrates the effectiveness of the analytical model for capacity evaluation.

5.6 Review Exercises

1. Derive the circuit data capacity expression in equation (5-49) for reverse link.
2. Calculate the mixed voice and data capacity numerically using equation (5-52) for reverse link.
3. Calculate the mixed voice and data capacity for forward link using equation (5-68) numerically.
4. In system level simulations, place N mobiles uniformly among M cell/sectors. Prove that when N is large and N/M is finite, the number of mobiles in the center cell is approximated to a Poisson random variable.

5.7 References

[Ber87] D. Bertsekas, R. Gallager, *Data Networks,* Prentice Hall, Inc., Englewood Cliff, New Jersey, 1987.

[Gro98] D. Gross, C. M. Harris, *Fundamentals of Queueing Theory, Third Edition,* Wiley Interscience Publication, 1998.

[Pap91] A. Papoulis, *Probability, Random Variables, and Stochastic Processes, Third Edition,* WCB McGraw-Hill Publication, 1991.

[Vit95] A. J. Viterbi, *CDMA: Principles of Spread Spectrum Communication,* Addison-Wesley Wireless Communications Series, 1995.

[Yac93] M. D. Yacoub, *Foundations of Mobile Radio Engineering,* CRC Press, 1993.

[Zha1_01] Q. Zhang, O. Yue, "UMTS Air Interface Voice/Data Capacity – Part 1: Reverse Link Analysis," *Proceedings of VTC2001 Spring,* May 2001.

[Zha2_01] Q. Zhang, "UMTS Air Interface Voice/Data Capacity – Part 2: Forward Link Analysis," *Proceedings of VTC2001 Spring,* May 2001.

[Sch92] S. C. Schwartz, Y. S. Yeh, "On the Distribution Function and Moments of Power Sums With Log-Normal Components," *The Bell System Technical Journal, Vol. 61, No. 7, pp. 1441-1462, September 1992.*

[25.201] 3GPP TR 25.201, "Physical Layer – General Description."

[25.211] 3GPP TR 25.211, "Physical Channels and Mapping of Transport Channels onto Physical Channels (FDD)."

[25.212] 3GPP TR 25.212, "Multiplexing and Channel Coding (FDD)".
[25.213] 3GPP TR 25.213, "Spreading and Modulation (FDD)".
[25.214] 3GPP TR 25.214, "Physical Layer Procedures (FDD)".
[25.215] 3GPP TR 25.215, "Physical Layer – Measurements (FDD)".
[3GPP2] 3GPP2 TSG-C.R1002, "1xEVDV Evaluation Methodologies".

Chapter 6

DESIGN AND TRAFFIC ENGINEERING OF A BASE STATION

6. BASE STATION DESIGN

As shown in Figure 4-5 in Chapter 4, a radio access network consists of one or multiple base station controllers and tens/hundreds of base stations connected together via the backhaul network. In this chapter, we describe how a base station (or NodeB in UMTS terminology) can be designed to meet certain performance requirements. First, we discuss how the CPU budgets for various cards within a UMTS base station can be determined. Then, we discuss how the interface bandwidth between a base station and its associated base station controller can be estimated. Later, similar design issues for a CDMA2000 base station are discussed. Techniques for processor performance enhancements are also discussed.

Figure 6-1 illustrates a simple high-level architecture of a base station. A base station typically consists of several key components:
- Various line cards that interface to the base station controller
- A main controller that processes the radio signaling messages, transport signaling messages, and performs resource management.
- Various radio cards that house channel elements/radios that allow the base station to send/receive radio signals to/from the mobile terminals. Each radio card may be designed to support a certain number of voice calls and/or data calls, e.g., a typical radio card with 32 channel elements can support 64 voice calls or 32 64–Kbps data calls.

Figure 6-1 also depicts both the bearer and signaling paths through the key components of a base station. The bearer traffic passes from the line card on the main controller board to one of the radio cards that house the channel element that has been assigned to process the bearer traffic. The signaling traffic passes from the line card to the main controller. After being

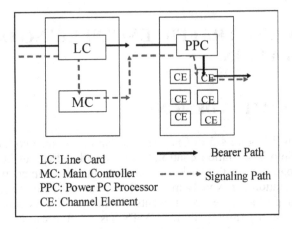

Figure 6-1. Logical Architecture for a Typical Base Station

processed, it is being sent to the assigned channel element in one of the radio cards to be sent to the mobile terminals. Multiple resources are required at a base station to support the voice and data users, e.g., power resources, CPU resources, interface bandwidth, and memory. The radio resources such as transmission power, and codes), and interference level determine the air interface capacity as discussed in Chapter 5. Given the air interface capacity, we first discuss how the CPU, and the interface bandwidth requirements for a UMTS base station (or NodeB) can be determined in Section 6.1. Then, we discuss similar design issues for the CDMA 1xEV-DO base station in Section 6.2.

6.1   UMTS Base Station Design

Here, we first describe how the CPU budgets for the various component cards within a UMTS base station (NodeB) are determined. Then, we discuss the factors that affect the interface bandwidth requirement and present some guidelines on the voice and data capacity that can be supported by a certain link bandwidth.

6.1.1   CPU Budget for Various Component Cards in NodeB

Before we can discuss how the CPU budget for the main controller, the line card, and the radio card can be determined, we need to find out how many voice and data users can be supported in the air interface per cell.

6.1.1.1 Air Interface Capacity

Using the approach described in Chapter 5, one can determine the air interface capacity based on the Quality of Service (QoS) requirement and the available RF resources. For example, a typical UMTS cell may be able to support an average data throughput, Ψ, 400 Kbps, or 104 voice legs per carrier.

The effective total available air link bandwidth, Ψ_{eff}, is defined to be the useful air link bandwidth available for the users without including the bandwidth used for supporting soft/softer handoff legs. Let us assume that the soft handoff rate is 30%. Then, $\Psi_{eff} = \Psi /1.3$.

The effective number of data users, each with a packet service rate of R, that can be supported by such a base station can be determined from the following equation:

$$N_{user} = \psi_{eff} / \left(R \times \left(1 - P_B \right) \right) \tag{6-1}$$

where P_B, the block error rate, is typically set at 10% for packet data users.

Next, we discuss how the CPU budget for the main controller, the line card and the radio card can be determined based on such air interface capacity numbers.

6.1.1.2 CPU Budget for the Main Controller

The Iub interface provides the service of exchanging information between the NodeB and RNC. Three types of messages: ALCAP, NBAP[25.427], and OA&M are carried over the Iub interface and have to be processed by the Main Controller. ALCAP protocol defines a set of messages exchanged between the NodeB and the RNC to set up, maintain, or release the Iub data transport bearer. Once the transport bearer setup is completed, signaling messages can be exchanged. The NBAP protocol defines the signaling procedures for the voice or data connections over the Iub link. Most importantly, NBAP is responsible for setting up and deletion of the radio links. Other tasks include management of common channels and common resources, measurement handling and control, etc. The message rates generated by NBAP, ALCAP are a function of key parameters such as soft handoff rate, number of calls, etc. In addition, signaling messages related to the radio resource control (RRC) procedures (between the user equipment

(UE) and the RNC) and the radio access network application part (RANAP) procedures (between the user equipment and the RNC) are also transported over the Iub link.

To determine the CPU budget for the main controller, we need to evaluate how many messages need to be processed by the controller. In the following paragraphs, we use the mobile originated call setup and call release procedures as an example to illustrate how the number of signaling messages incurred can be estimated.

For each mobile originating call, the following procedures will be invoked:

1. Mobile Originated Call Setup Procedure
- RRC connection establishment for DCH establishment

Figure 6-2 shows the signaling messages that are exchanged for establishing a Radio Resource Control (RRC) connection during the process of establishing a dedicated channel (DCH). For DCH establishment, RRC connection request is sent by the User Equipment (UE) to the Serving RNC (SRNC) across the Uu and Iub interface. The SRNC then sends a Radio Link Setup Request message to the NodeB across the Iub interface. The NodeB responds with a Radio Link Setup Response message across the Iub interface. Then, the SRNC and the NodeB exchange the ALCAP Iub Establishment Request/Confirm messages. After that, the SRNC sends a Downlink Synchronization message to the NodeB and the NodeB responds with an Uplink Synchronization message. This is followed by a RRC Connection Setup Response message from the SRNC to the UE across both the Iub and Uu interfaces, and the UE sends a RRC Connection Setup Complete message to the SRNC.

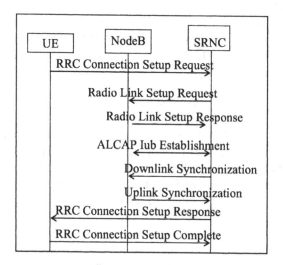

Figure 6-2. RRC Connection Establishment: DCH Establishment

- NAS signaling connection establishment

An initial direct transfer message containing the Connection Management Service Request will be sent from the UE to the SRNC. The SRNC then sends an initial UE message to the core network and the core network responds with a SCCP Call Control message.

- Common ID procedure

The Core Network (CN) will send a Common ID message to the SRNC.

- NAS authentication procedure

This consists of an authentication request message from the CN to the SRNC, which is relayed by the SRNC to the UE via a Downlink Direct Transfer message. Then, an authentication response message is sent from the UE to the SRNC and is relayed via a Direct Transfer message by the SRNC to the CN.

- Integrity protection procedure

This consists of a Security Mode Command message from the CN to the SRNC that the SRNC relays to the UE. The UE responds with a Security Mode Complete message to the SRNC.

- NAS call setup procedure

This consists of a Call Connection Setup message from the UE to the SRNC, and the SRNC will forward it to the CN. The CN will send a Call Proceeding message to the SRNC and the SRNC relays this message using a Downlink Direct Transfer message to the UE.

- RAB assignment procedure

Figure 6-3 shows the signaling messages that are exchanged to establish the radio access bearer (RAB). First, the CN sends a RAB Assignment Request (RANAP/SCCP message) to the SRNC. For a circuit domain request, the SRNC will send a Radio Link Reconfiguration (prepare DCH addition) message (which is an NBAP message) to the NodeB. The NodeB responds with a Radio Link Reconfiguration Ready message (also an NBAP message). Then, the SRNC sends an ALCAP Iub Establish Request message to the NodeB, and the NodeB responds with an ALCAP Iub Establish Confirm message. Later, the SRNC sends a Downlink Synchronization message (which is a DCH Frame Protocol message) to the NodeB and the NodeB sends an Uplink Synchronization message (also a DCH Frame Protocol message) to the SRNC. The SRNC then sends a Radio Bearer Setup message (which is an RRC message) to the UE and the UE has to respond with a Radio Bearer Setup Complete message (also an RRC message). This is followed by an RAB Assignment Response message (RANAP/SCCP message) from the SRNC to the Core Network.

- ALCAP Iu data transport bearer setup procedure

This consists of an ALCAP Iu Establish Request message sent from the SRNC to the CN and an ALCAP Iu Establish Confirm message sent by the CN to the SRNC.

- NAS alerting/connect procedure

This consists of an Alerting message sent from the CN to the SRNC, which the SRNC relays to the UE via a Downlink Direct Transfer message. The CN also sends a Connect message to the SRNC that is relayed by the SRNC to the UE via a Downlink Direct Transfer message and the UE sends a Connect Acknowledge message to the SRNC that is relayed to the CN.

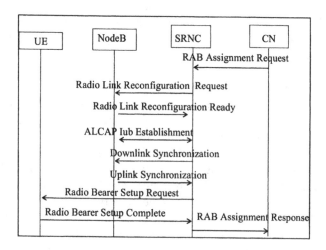

Figure 6-3. RAB Assignment Procedures

2. Mobile Initiated Call Release Procedure

- NAS call release – mobile initiated

This consists of a Disconnect message (which is an RRC message) sent by the UE to the SRNC that the SRNC relays to the CN via the Direct Transfer message (RANAP+SCCP+CC). The CN then sends a Release message to the SRNC that the SRNC relays to the UE via a Downlink Direct Transfer message (RRC message). The UE then responds with a Release Complete message (RRC message) to the SRNC that is relayed to the CN.

- RAB release procedure

The CN sends a RAB Assignment Request (release) message to the SRNC and the SRNC responds with a RAB Assignment Response message to the CN. The SRNC then sends a Radio Link Reconfiguration Prepare (DCH deletion) message (NBAP message) to the NodeB and the NodeB responds with a Radio Link Reconfiguration Ready message (NBAP message). The SRNC then sends a Radio Link Reconfiguration Commit message (NBAP message) to the NodeB and also a Radio Bearer Release message to the UE (RRC message). The UE responds with a Radio Bearer Release Complete message. The SRNC then sends an ALCAP Iub Request Request message to the NodeB and the NodeB responds with an ALCAP Iub Release Confirm message.

- ALCAP Iu data transport bearer release procedure

This consists of an ALCAP Iu Release Request message from the SRNC to the CN and an ALCAP Iu Release Confirm message from the CN to the SRNC.

- RRC connection release procedure

This consists of an Iu Release Command from the CN to the SRNC and an Iu Release Complete message from the SRNC to the CN, a SCCP Release message from the CN to the SRNC and a SCCP Release Confirm message from the SRNC to the CN. The SRNC then sends a RRC Connection Release message to the UE and the UE responds with a RRC Connection Release Complete message. The SRNC then sends a Radio Link Deletion Request message to the NodeB, and the NodeB responds with a Radio Link Deletion Response message. This is followed by an ALCAP Iub Release Request message from the SRNC to the NodeB and the NodeB responds with an ALCAP Iub Release Confirm message.

For each soft handoff event that needs to be performed, two additional NBAP messages (radio link setup request/response) need to be exchanged between an RNC and the NodeB. Two ALCAP Iub data transport bearer setup messages, a downlink and uplink synchronization message (DCH–Frame Protocol messages), an RRC active set update command, and a complete message need to be sent between the SRNC and the UE. In addition, there are possibly two NBAP messages for radio link setup deletion request/response, and two for ALCAP Iub data transport bearer release messages. Therefore, depending on the assumption of the number of soft handoff events per voice call duration, we add accordingly these extra messages to the total number of NBAP, ALCAP, and RRC messages that need to be sent per mobile originated call.

By counting the number of messages across each interface and the total signaling bytes per interface, the numbers of signaling messages generated during each Mobile Originated (MO) call are summarized in Table 6-1.

Table 6-1 Number of Signaling Messages Per Mobile Initiated Call Origination

Protocol	Number of DL msgs	Avg DL Msg Size (byte)	Number of UL msgs	Avg UL Msg Size (byte)
NBAP	6	139	4	73
ALCAP	4	39	4	11
Iu RANAP	12	89	11	109

Similar counting can be done for other call scenarios, e.g., Mobile Terminated (MT) call, location update procedure, routing area/URA update procedures, IMSI attach/detach procedures, URA_PCH to Cell_DCH transition, Cell_DCH transition to URA_PCH transition due to the RRC inactivity timer expiry, PS attach & PDP activation, PDP deactivation and PS detach. Then, the total CPU budget for the main controller can be estimated.

The following assumptions are made to estimate the CPU budget for the main controller:
- There are 3 cells per NodeB.
- There is no packet data traffic.
- The RNC is designed to support 1000 Erlangs.
- The number of IMSI attach procedure per mobile per Busy Hour (BH) is 1.4
- The number of IMSI detach procedure per mobile per BH is 0.1
- The number of PS attach and PDP activation per BH is 0.1
- The number of PS detach and PDP activation per BH is 0.1
- The number of Routing Area (RA) Update per BH is 0.06
- The number of URA Updates per BH is 1
- The number of soft handover legs is 1.3
- The number of soft and softer handover legs is 1.5
- The voice traffic rate per call is 12.2 Kbps.
- The circuit voice activity factor is 50%
- The average voice call duration, Voice_Call_Dur, is 120 seconds.
- The proportion of Mobile Originated (MO) and Mobile Terminated (MT) calls is 50%, 50% respectively.
- The average size of MO or MT call messages is 7 bytes.
- The probability of invoking authentication procedure per MO or MT call is 0.5
- Every MO or MT call invokes integrity protection procedure.
- Number of Busy Hour Voice Call Attempt (BHCA) per subscriber is 0.6

We further assume that the airlink capacity evaluation reveals that we can support 104 voice legs per cell with a Block Error Rate of 1%. With the assumption that the number of soft and softer handover legs is 1.6, we can support $104/1.6 = 65$ Dedicated Channel (DCH) pipes and ($65 \cdot 1.3 = 84.5$) DCH pipes on the Iub link (since softer handoffs will not have any effect on the Iub link).

The total number of voice subscribers per cell = 3240
The total voice calls per busy hour = $3240 \cdot 0.6 = 1944$

With these numbers, we can determine the total number of messages, and the total message sizes for each procedure.

For example: for each voice MO call, there are 4 NBAP uplink messages with an average message size of 73 bytes (without transport network overhead). After adding the overhead due to AAL5 trailer, SSCF–UNI, SSCOP, ATM/SDH layers, the average message size is 110 bytes. So the total number of bytes per call is 110·4 = 440 bytes. Thus, the total signaling bandwidth to support NBAP uplink message from all voice subscribers is:

440·8·1944/3600/1000 Kbps = 0.95 Kbps

Similarly, for each voice MO call, there are 6 NBAP downlink messages with an average message size of 138 bytes (without transport network overhead). After adding the overhead due to AAL5 trailer, SSCF–UNI, SSCOP, ATM/SDH layers, the average message size is 220 bytes (note that this number is higher because more ATM cells are required to carry 138–byte payload). Thus, the total number of bytes per call = 220·6 = 1320 bytes.

Thus, the total signaling bandwidth required to support the NBAP downlink message from all voice subscribers is:

1320·8·1944/3600/1000 Kbps = 2.85 Kbps

By doing such estimation on the number of messages that are sent in each major NBAP and ALCAP procedure per user, one can easily come up with a table as shown in Table 6-2. Each column of the 4th row in Table 6-2 is the sum of the corresponding columns in the 2nd and 3rd row. Similarly, each column in the 9th row in Table 6-2 is the sum of the corresponding columns from rows 5 to 8. Each column in the final row is the sum of the corresponding columns from rows 4, 9, and 10.

From Table 6-2, we see that the total NBAP and ALCAP message rates per NodeB is about 45 messages per second (4th row, 1st column) with a very low soft handoff rate of one per call and about 104 messages per second (4th row, 4th column) with a high soft handoff rate of 15 per call. The soft handoff rate per call is defined as the number of soft handoffs requested during the duration of the call. The message rates generated by data users are smaller than that for voice users. With the numbers listed in Table 6-2, one can then find a processor for the Main Controller board with enough processing power to process the total messages generated by both voice and/or data users supported by a NodeB at full capacity.

Table 6-2: Message Rate per Second Generated by Voice Users as a Function of Soft Handoff Rate/Call

Soft handoff rate	1	5	10	15
Call related NBAP	24	32	42	52
ALCAP	21	29	39	49
Call related NBAP+ALCAP	45	61	81	101
Common measurement report	10	10	10	10
System info update	10	10	10	10
Ded. measurement report	3	3	3	3
Ded. measurement init. and term.	1.2	2.6	4.4	6.1
Common and ded. measurements	24.2	25.6	27.4	29.1
OA&M	3.2	3.2	3.2	3.2
Total msgs/second	72	90	113	136

6.1.1.3 CPU Budget for the Line Card

The line card at a NodeB terminates the ATM connection from the RNC. From the air interface analysis, one can derive the number of simultaneous voice calls, the number of simultaneous 64 Kbps and/or 384 Kbps data sessions that can be supported per cell/carrier.

Consider the following simple example: assume that the NodeB needs to support

(a) 81 voice calls where the voice activity factor is 50%

Within 10 ms, each voice packet generates about 1 ATM cell. Thus, the NodeB needs to support a total of 40.5 ATM cells every 10 ms for the voice scenario. This means that the line card needs to process one ATM cell every 192 µs for the voice scenario.

(b) 40 64–Kbps users

Assume that the transmission time interval (TTI) of each 64–Kbps user is set to 10 ms. This means that within 10 ms, each user generates 80 bytes data which can only be fit into the payloads of two ATM cells. Therefore, within 10 ms, each 64–Kbps user generates two ATM cells. For 40 such users, a NodeB needs to support a total of 80 ATM cells every 10 ms. This means that the line card needs to process one ATM cell every 125 µs for the 64–Kbps user scenario.

(c) 6 384–Kbps users.

Again, assume that the TTI of each 384–Kbps user is set to 10 ms. Within 10 ms, each Iub frame from a 384–Kbps user generates about 12 ATM cells. Thus, the NodeB needs to support a total of 72 ATM cells per 10 ms. This means that the line card needs to process one ATM cell every 140 µs for the 384–Kbps user scenario.

In addition to the bearer traffic, the line card also needs to route the signaling messages, e.g., NBAP and ALCAP messages. The signaling message rate and traffic generated from voice and data users can be determined using the same methodology described in Section 6.1.1.1.

Considering the same voice example as before: assume that 81 voice users generate about 50 Kbps of signaling traffic using an assumption that each call results in 5 handoffs. We include another 30% overhead for other unaccounted procedures, thus a total of 65 Kbps signaling traffic needs to be processed. This is equivalent to about 1.4 ATMs every 10 ms. So, taking the voice users scenario, each line card needs to process about $(40.5 + 1.4) \sim 42$ ATM cells every 10 ms. For both signaling and bearer traffic, if we want the processor in the line card to operate at 70% utilization, then it means that the processor has to process one ATM cell in 130 µs. Using such CPU budgets, one can then search for the right type of processor and its associated clock rate to process the total message (bearer + signaling) rate generated by all users at full capacity.

6.1.1.4 Radio Card CPU Budget

The major component of the radio card is a processor and a set of channel elements. The main functions of the processor is to read the data packets from the bus into the processor, perform frame protocol functions, and then write the processed data to the appropriate channel element.

The effective average traffic generated by one AMR voice call including frame protocol overhead is about 16.4 Kbps. This traffic is transported via the Firewire bus to the processor in the radio card. Including the Firewire overhead, each voice call generates about 66 bytes every 20 ms. The Firewire bus has a maximum throughput of about 100 Mbps. So, the bus controller takes about $66 \cdot 8 \cdot 50 / (100 \cdot 1000) = 0.264$ to process traffic from a voice user. Assuming that each radio card supports a maximum of 32 voice calls. At 100% processor occupancy, each voice call needs to be processed within about 0.6 ms. Assuming 50% processor occupancy, then the radio card framing protocol and channel element scheduling functions for a voice call must be completed within 0.3 ms.

The effective average traffic generated by one connected 64 Kbps packet data user including frame protocol overhead is about 70 Kbps. With some overhead for the bus (e.g., Firewire), this translates to about 198 bytes per 20 ms. For 384 Kbps, the corresponding number is about 534 bytes per 10 ms frame.

Each radio card can support a maximum of 16 64–Kbps data calls or 6 384–Kbps data calls. Thus, to process 16 64–Kbps users within 20 ms (or 6 384–Kbps users within 10 ms), the frame protocol and channel element scheduling functions need to be completed within 1.2 ms (or 1.58 ms for the 384–Kbps case) at 100% processor occupancy. For 50% processor occupancy, the functions need to be completed within 0.6 ms and 0.79 ms for 64 Kbps and 384 Kbps correspondingly. Again, such numbers help us to determine which processor type to use for the radio card housing the channel elements and how many channel elements can be supported per card.

6.1.2 Iub Interface Capacity

Tens or hundreds of base stations (or NodeBs) are connected to the radio network controller (RNC) via a backhaul network. The logical interface between a NodeB and the RNC is referred to as the Iub interface while the logical interface between two RNCs is referred to as the Iur interface. The protocol stack for a dedicated channel service between the RNC and NodeB is shown in Figure 6-4. Application data is passed to the RLC/MAC layer where it is segmented and appropriate headers are inserted. Then, it is passed to the Dedicated Channel (DCH) Frame Protocol (FP) layer. The DCH FP layer defines the structures of the data frames carrying user data across the Iub interface. In addition, DCH FP also defines the control frames used to exchange control and measurement information across the Iub interface. The DCH FP frames are transported via AAL2 connections between the RNC and the NodeB.

It is important to understand how the link bandwidths for these interfaces are determined. Before we can answer this question, let us explore the various factors that affect how much bandwidth should be allocated at these interfaces.

For any active bearer channel, the RNC produces a frame for transmission on the air link once every transmission time interval (TTI). Each frame is

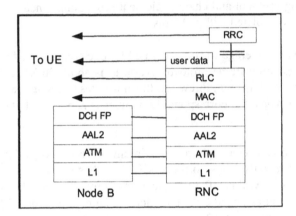

Figure 6-4. Protocol Stack at the RNC and NodeB

indexed by a connection frame number (CFN). The CFN is incremented every TTI. The choice of a TTI depends on the Quality of Service (QoS) requirement of the application. For a delay sensitive service such as circuit voice, TTI is often chosen to be either 10 ms or 20 ms while for a delay tolerant service like web-browsing, larger TTI values (either 40 ms or 80 ms) can be used.

6.1.2.1 Voice Service

Since voice has a tight delay requirement, let us see how the frames generated from the transcoder are sent from the core network element (such as MSC) to a User Equipment (UE). Figure 6-5 [Abr02] shows the flow of frames from the Core Network to a UE and their associated delays.

The total delay time between the when the transcoder produces a frame, to the time NodeB sends it out on the air interface is $(T_5 - T_0)$. The various components contributing to the delay are shown in Figure 6-5. One can control the statistics of the delays by using appropriate engineering guidelines for the links to ensure that the bandwidth utilization is neither too high (excess delay jitter) nor too low (inefficient).

A timing adjustment algorithm is provided to dynamically adapt the RNC transmission time, T_3 to the Iub interface link delay $(T_4 - T_3)$. The RNC gets feedback about the link delay conditions and adjusts its transmission time accordingly. The main goal is to maximize the probability that the frame reaches the NodeB in time to be sent out on the air interface at its scheduled

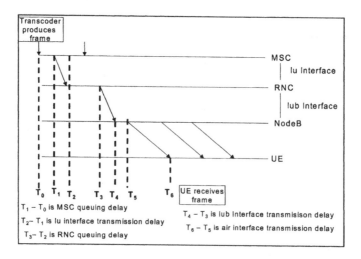

Figure 6-5. Flow of Frames from the Core Network to a UE

time. To limit the buffer requirements at the NodeB, it is also desirable that the frame does not arrive too early. The 3GPP standards provide a framework for performing timing adjust [25.402]. The 3GPP standard defines a receive window at the Node B as shown in Figure 6-6.

TOAWE and TOAWS are parameters that can be set. The Time Of Arrival (TOA) is measured from the edge of the receive window (shaded region in the figure) as shown in the figure. Positive TOA implies the frame arrived before the right edge of the window while a negative TOA implies that the frame arrived after the right edge of the window. If the frame arrives after Latest Time Of Arrival (LTOA), then the NodeB discards the frame. The resolution and range of the parameters TOAWE, TOAWS, and TOA are provided in [25.402].

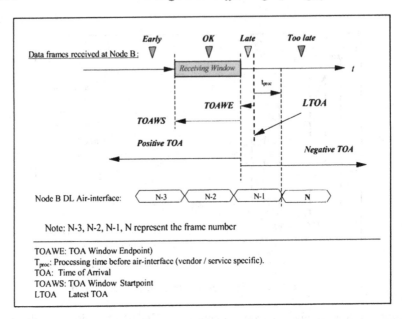

Figure 6-6. Timing Adjust Receive Window

When a frame is received outside the "receive window," the NodeB generates an "UL Timing Adjust Frame" and sends it to the RNC. Based on the TOA information received from the NodeB, the RNC will appropriately adjust the transmission time for the next frame scheduled to be sent to the NodeB.

Given a desirable frame discard probability, there is a maximum number of voice or data users (assuming a certain allocated rate per voice/data user) that can be supported by a certain link bandwidth at the Iub interface. We refer to this maximum number as the Iub link capacity. Since the timing adjustment influences how many frame protocol frames will arrive within the received window, it is clear that the Iub link capacity is dependent on how timing adjustment is performed. One possible timing adjust algorithm is described in [Abr02].

In the absence of the UL timing adjust frame,

$$t_{N+1} = t_N + t_{TTI} \qquad (6\text{-}2)$$

If an UL timing adjust frame is received, then a fixed step adjustment of size delta is made. If the TOA indicates that the frame arrived to the "left" of the receive window, then the transmit time of the next frame is delayed, i.e.

$$t_{N+1} = t_N + t_{TTI} + \delta \qquad (6\text{-}3)$$

If the TOA indicates that the frame arrived to the "right" of the receive window, then the transmit time of the next frame is advanced, i.e.

$$t_{N+1} = t_N + t_{TTI} - \delta \qquad (6\text{-}4)$$

Some simulation studies of this algorithm were carried out in [Abr02] to determine the call carrying capacity for the Iub interface. The capacity of the Iub link is defined to be the maximum number of users for which the frame discard probability is no more than 0.05%. In their work, the AMR speech activity was simulated as a Markov process with an average on time of 350 ms and an average off time of 650 ms. Both the on and off times are assumed to be exponentially distributed. TOAWS and TOAWE are set to 5 ms and 1 ms respectively. The results are tabulated in Table 6-3.

Table 6-3 Voice Call Carrying Capacity for Different Interface Link Rates

Interface link rate	Capacity (number of users)	UTRAN delay	Measured link occupancy	Statistical multiplexing gain
768 Kbps	55	10.56 ms	80.20%	29.30%
1.5Mbps (T1)	118	10.50 ms	86.00%	41.10%
3.072Mbps	249	10.05 ms	92.90%	62.50%
4.068Mbps	382	10.06 ms	93.00%	66.30%

In Table 6-3, the mean UTRAN delay is defined to be the time between when the frame was available at the Frame Protocol (FP) layer at the RNC and the scheduled transmit time from the NodeB to the UE. The statistical multiplexing gain is defined as the percentage increase in the number of calls that can be supported with activity detection over the number of calls that can be supported without activity detection (assume peak rate). The FP and AAL2/ATM overheads are considered in the reported voice capacity numbers. Even though the voice activity is less than 50%, the statistical multiplexing gain is only between 29% to 66% for the bandwidth considered due to the stringent delay requirement imposed. The statistical multiplexing gain becomes worse with packet data service. We elaborate on this in the next subsection.

6.1.2.2 Packet Data Service

The data generated by a user from the network is mapped into Iub frames when the data arrives at RNC. These frames are then sent on the Iub-link. The mapping has to conform to the format specified in the 3GPP documentation [25.101]. Once the TTI and the Radio Bearer Rate (ranging from 64 Kbps to 384 Kbps for Web-browsing interactive service) are given, the Transport Format Set can be determined from [25.101]. The RLC layer will fragment the data packet into appropriate RLC protocol data units (PDUs) and pass them to the Medium Access Control (MAC) layer. The MAC layer adds the appropriate headers and passes them to the dedicated channel framing protocol (DCHFP) layer. The DCHFP layer adds the framing protocol headers to the frames. The framing protocol PDU is then passed to the ATM adaptation layer (AAL2). These AAL2 layer frames from different users are then packed into ATM cells before being transmitted on the Iub-link (see the protocol stack shown in Figure 6-4). Different facility link bandwidths may be considered, e.g., T1, E1 etc.

Packet data traffic tends to be bursty, i.e., there are periods with very high traffic and periods with very low traffic. During periods of high traffic, incoming packets may have to be queued causing delays, packet losses, and inefficient usage of the link bandwidth. For both the CDMA 3G1x and the UMTS R99 systems, the base station expects the MAC protocol data units to arrive within a certain time window. If the packets arrive late, the packets will be discarded.

Each packet data service belongs to one of the 4 QoS classes specified in the 3GPP standards: conversational, interactive, streaming and background. Table 6-4 shows a typical parameter set for the transport channel for the interactive class as specified in [25.427]. The transport format set (TFS) contains five different transport formats or sizes (TF0–TF4). If transport format TF1 is chosen, one 42-byte transport block is sent every TTI (of 20 ms). If the largest transport format TF4 is selected, then four 42-byte transport blocks are sent per TTI, giving the user a peak rate of 64 Kbps. Suppose the RLC buffer holds a 1500-byte packet. One fragmentation option would be that the MAC creates 9 TF4's (each with size 160 bytes), which would hold 1440 bytes of user data. The remaining 60 bytes would be placed in the smallest TFS block that would hold it, namely TF2 (size 80 bytes).

To provide reasonable bandwidth utilization on the Iub link, statistical multiplexing of data sources is necessary. However, with statistical multiplexing, there will be time durations when the offered load temporarily exceeds the link bandwidth. For example, with overheads, the peak rate of a 64 Kbps user is around 82.5 Kbps. For a T1 link (1536 Kbps), the number of users that can be supported is 1536/8.25 = 18. If the number of users with non-empty RLC buffers exceeds 18 at any given time, then the input rate to the Iub link is more than what it can handle. This may result in excessive frame discards at the NodeB.

Some simulation studies on the session carrying capacity of an Iub link for packet data users running Web-browsing sessions were reported in [Sar03a], [Sar03b]. Results reported in [Sar03a] indicate that the useful link utilization is a function of the peak airlink data rate, the traffic model etc. In [Sar03a], the authors reported the Iub link utilization using 0.5% as the desirable frame discard probability. It was reported that for a T1 link, the allowable link utilization to support 64, 128 and 384 Kbps users while meeting the desirable frame discard probability is 57%, 43%, and 14% respectively assuming a certain traffic model for the web-browsing users. In Chapter 7, we describe a rate control technique that helps to improve the link utilization when data users with high bit rates are supported at the expense of a slight degradation of the peak rate allocated to the data users.

Table 6-4 Transport Format Set for the Interactive/Background Service Type for the 64 Kbps Radio Bearer

Transport channel parameters for Interactive or background for the 64–Kbps PS RAB		
RLC	Logical channel type	DTCH
	RLC mode	AM
	Payload sizes (byte)	40
	Max data rate (Kbps)	64
	RLC header (byte)	2
MAC	MAC header (byte)	0
	MAC multiplexing	N/A
Layer 1	TrCH type	DCH
	TB sizes, bytes	42
	TFS TF0 (byte)	0x42
	TF1 (byte)	1x42
	TF2 (byte)	2x42
	TF3 (byte)	3x42
	TF4 (byte)	4x42
	TTI (ms)	20

6.2   Capacity Evaluation and Resource Management of 1xEV-DO Base Stations

In this section, we first give a high-level overview of the 1xEV-DO base station. Then, we describe a methodology for analyzing and estimating the processor occupancy of the cards used in the base station. This is followed by some discussions of the processor performance enhancements.

6.2.1   1xEV-DO Base Station Architecture

The 1xEV-DO system utilizes a complete CDMA2000 carrier to support the high-speed data only traffic. The base station architecture is similar to the 3G1x base station as shown in Figure 6-7. In the actual network deployment, 1xEV-DO base station usually coexists with the 3G1x base station but operates at a different carrier to support the DO traffic. The base station consists of three main parts:

- CDMA Radio Controller
- 1xEV-DO Modem boards
- Base-band radio and peripherals

CDMA Radio Controller (CRC) mainly routes the traffic for both forward link and reverse link between the 1xEV controller and the base station, specifically the modem boards. In addition to routing bearer and signaling packets, CRC also handles base station OA&M (Operation, Administration, and Management) processing. The backhaul (i.e., T1/E1) interface between the cell and the 1xEV controller is terminated at the CRC.

The 1xEV modem board (EVM) terminates the air interface and handles all RF (Radio Frequency) signaling processing. Depending on the ASIC architecture and design, a separated forward link modem and reverse link modem card each with its own IP address can be used to process the transmitting and receiving of the air interface packets in the forward and reverse direction, respectively. The interface between the modem board and the CRC is usually a high-speed bus.

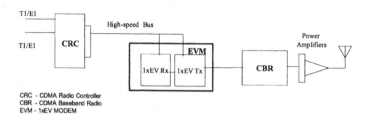

Figure 6-7. CDMA EV-DO Base Station Architecture

CDMA base-band radio (CBR) usually has three phases that handle the radio part of the three sectors. It works together with other components, such as a timing/frequency unit and an amplifier, to send the signal out on each carrier.

The critical resources at the base station are the real-time processing of each component (multiple CPUs in the CRC, the modem boards, etc.) and the high-speed buses connecting the components. It is important to do capacity planning and resource management so that we can guarantee that the base station can handle the high-speed data and signaling traffic effectively.

Performance and capacity models are essential in evaluating the capacity of each component and identifying the resource bottlenecks at the base station. Performance modeling has collateral benefits as well. It enables us to project the effect of new features on critical resources. It helps us to project what the system can offer and reduce the unplanned field-testing. It also provides useful guidance to future product platform evolution.

6.2.2 Processor Occupancy Analysis

One of the critical resources is the real-time processing of the processor, for example, the CPU of the CRC components. Processor Occupancy (PO) is defined to be the fraction of time the CPU spends on time critical tasks. The estimation should include only the real-time tasks, such as the VxWorks real time tasks. The total CPU utilization includes other non-real-time tasks and no load occupancy. Here, it is important to note that the PO for time critical tasks is the main indicator of the performance measure. The total CPU utilization provides useful information but does not represent the critical resource utilization.

The basic method of estimating processor occupancy for a given traffic activity is as follows.

$$P = R \times T \tag{6-5}$$

where P is the processor occupancy; R is the rate at which the traffic is generated; and T is the time spent by the processor to process the traffic activity.

For example, a data throughput of 800 Kbps or 100 Kbytes/second results in 1000 packets/second with packet size of 100 bytes. Each 100-byte packet consumes 0.8 ms at the processor. As a result, the processor occupancy is $P = 1000 \times 0.8 / 1000 = 80\%$.

The first step to estimate the PO is to identify the traffic activities at the processor. There are usually two types of traffic that need real-time processing, i.e., bearer traffic and associated signaling traffic (e.g., call setup messages). The second step is to estimate the rate generated by each traffic activity. This requires careful attentions to system implementation details and understanding of the traffic functionality. The third step is to obtain the packet processing time. Detailed and accurate measurement is essential here. In addition, it also requires understanding of the measurement setup and packet transit path to analyze the measurement data. Packet processing time usually depends on the packet length, the type of packets, and the transit direction (forward or reverse direction.) In the final PO estimation, all the above factors need to be taken into consideration.

We give a simple example to illustrate the process of evaluating the PO.

Example:
Estimate the processor occupancy of a processor (i.e., CRC) for real-time processing at the EV-DO base station.

Step 1: Identify the traffic activities at the processor.
The processor at the CRC mainly routes the packets between the base station controller and the modem boards in both forward and reverse directions. It does protocol conversion and simple IP protocol processing (IP packet routing and fragmentation). Data bearer and signaling/control traffic messages pass through the processor. The message unit is IP packet with variable length and packet size.

Step 2: Estimate the rates generated by the traffic activities.

Suppose the processor handles the traffic of three sectors in one EV-DO carrier. We estimate the rates generated by traffic activities on forward link and reverse link separately.

Estimation on the forward link:

Assume that the capacity per sector on the forward link is about 600 Kbps at the RLP (Radio Link Protocol) layer. Given each RLP packet of 128 bytes, the rate of the RLP packets is

$$R_{RLP} = \frac{C_{thput}}{S_{pkt}} = \frac{3 \times 600 \times 1000}{128 \times 8} \approx 1757.8 \qquad (6\text{-}6)$$

where R_{RLP} is the number of RLP packet per second; C_{thput} is the throughput per cell; S_{pkt} is the RLP packet size.

The RLP packets are encapsulated into an IP packet at the base station controller, and sent over to the processor at the CRC. The rate of those IP packets is of much interest in estimating the PO. Each IP packet may consist of multiple RLP packets depending on the packet aggregation scheme and implementation. In order to calculate the rate, we need to determine the number of RLP packets per IP packet. For simplicity, let us assume that one IP packet may contain 1, 2, and up to 10 RLP packets with equal probability. Therefore the average number of RLP packets per IP packet is 5. The rate of the IP packets is the rate of the RLP packets divided by 5, which is

$$R_{IP} = \frac{R_{RLP}}{N_{RLP}} = \frac{1757.8}{5} \approx 351.6 \qquad (6\text{-}7)$$

where R_{IP} is the number of IP packets per second; N_{RLP} is the average number of RLP packets per IP packet.

Obviously the size of the IP packet also depends on the number of encapsulated RLP packets and the additional protocol headers. According to the protocol stack between the base station controller and the base station in the EV-DO RAN, the protocol stacks for bearer traffic from the RLP layer are: RLP – RMI – UDP – IP, where RMI stands for Remote Method Interface, which is a proprietary client-server interface for communication between the controller and the base station. The corresponding protocol overhead over the RLP packets is the summation of RMI header (19 bytes), UDP header (8 bytes), and IP header (20 bytes). The total protocol overhead is thus 47 bytes. Since an IP packet can encapsulate 1 up to 10 RLP packets, the IP packets sent over to the processor at the CRC in the forward direction will have 10 different sizes. Considering each RLP packet size of 128 bytes, the resulting IP packet will have 10 different sizes, varying from 175 bytes up to 1327 bytes. The IP packet size is an important parameter in determining the packet processing time, as described in the next step.

Estimation on the reverse link:

In EV-DO reverse link, there are multiple physical layer channels supporting different data rates. The supported data rate set is 9.6, 38.4, 76.8, 153.6 Kbps. The transmission rate of each mobile is controlled by the base station. The frame duration for each physical channel is fixed at 26.67 ms. Therefore, when a mobile is actively transmitting data, the rate generated by the physical layer is 37.5 frames/second. Each physical layer frame is processed by the modem card and a corresponding IP packet is constructed. Suppose that an IP packet consists of only one MAC packet from one mobile. The aggregate rate of the IP packets is based on the simultaneous number of active users, which is subject to the air interface capacity constraint.

$$R_{IP} = 37.5 \times N_{user} \qquad\qquad (6\text{-}8)$$

where N_{user} is the number of simultaneous users in transmission per cell.

Let us assume that the maximum number of simultaneous mobiles supported is 20 per sector per carrier. This corresponds to the worst case in terms of the number of IP packets per second. The rate of IP packets generated in a three-sector cell is thus $R_{IP} = 37.5 \times 20 \times 3 = 2250$

The frame size varies based on the data rate of each physical channel. For example, for a 9.6 Kbps reverse link channel, the corresponding frame size is 32 bytes. Adding 47-byte protocol overhead, the received IP packet size at the CRC is 79 bytes.

Step 3: Obtain the packet processing time.

Essential and accurate measurement is necessary in determining the packet processing time. The measurement setup is to connect the CRC processor between a server and a client. The server sends packets to the client through the processor. Each run consists of large number of packets with a fixed size. The throughput is measured at both the server and client.

The total CPU utilization is measured at the CRC processor. We shall treat the forward and reverse direction separately in the measurement and analysis.

Measurements on the forward link:

Measurement data of multiple runs with different packet size and throughput is collected and analyzed. The data are summarized and illustrated in Figure 6-8.

As shown in the figure, each set of measurement samples corresponds to the total processor occupancy versus the number of packets (with a fixed packet size) per second processed by the CPU. A linear curve fitting was performed on each set of the data and the linear equation is shown beside each line. To further interpret the curve, we can see that the intercept value at the y-axis reflects the no-load occupancy, which is approximately 2–3% of the CPU. The slope of the curve represents the packet processing time. With a proper translation, the packet processing time for the three different packet sizes is 3.221 ms, 0.881 ms, and 0.796 ms, respectively.

Apparently the packet processing time depends on the packet size. With more detail study on the software implementation of packet processing, we found that the packet processing time depends on the number of IP fragments generated by each packet. As we described earlier, the processor sends the IP packets to the modem board via a high-speed bus. The high-speed bus used in the base station is the Firewire Bus. The Firewire bus protocol and implementation utilizes a maximum payload unit (MTU) of 460 bytes. If the incoming IP packet is larger than 460 bytes, the IP packet will be fragmented so that each IP fragment can be fit in the payload of the Firewire packet carried on the bus. For the three different sizes used in the measurement, the 1428-byte packet results in 4 IP fragments, while the 458-byte and 228-byte packet still remains as one IP packet without any fragmentation. From the measurement data, it seems that the packet processing time is linear to the number of IP fragments. The processing time for the 458-byte and 228-byte packet is very close, around 0.8 ms. The 1428-byte packet processing time is approximately 4 times of the 458-byte or 228-byte packet processing time. As a result of the measurement analysis, we may draw the conclusion that the packet processing time is approximately 0.8 ms per IP fragment for this processor.

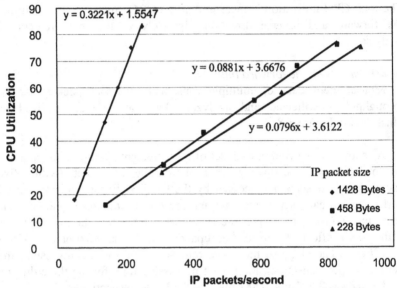

Figure 6-8. Processor Occupancy Measurement on Forward Link

Measurement on the reverse link:

Using the same measurement setup on the forward link, we measured the processor occupancy versus throughput on the reverse link. The results are shown in Figure 6-9.

Similar to the conclusion on the forward link, the packet processing time on the reverse link depends on the number of IP fragments generated by each IP packet. Furthermore, the packet processing time depends on the size of the IP fragment as well. As a result, we determine the packet processing time based on the actual measurement data for different packet size. The processing time varies from 0.66 ms to 1.1 ms.

Final step: Calculate the PO.

As we recall, the average number of IP packets per second on forward link in a 3-sector carrier is 2250 packets/second. Assuming that the processing time for each IP fragment is 0.8 ms, the IP packet processing time versus the packet size is summarized in Table 6-5.

The average packet processing time is 1.68 ms based on the uniform IP packet size distribution. Therefore the PO for the forward link bearer traffic is

$$P_{FL} = 351.6 \times 1.68 / 1000 = 59.07\% \qquad (6\text{-}9)$$

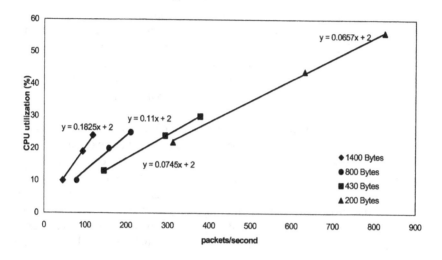

Figure 6-9. Processor Occupancy Measurement on the Reverse Link

The reverse link packet processing time varies from 0.66 ms to 1.1 ms. Even with maximum PO of 100%, the processor can only handle fewer than 1500 packets per second. As discussed earlier, the air interface capacity indicates that the number of IP packets per second on the reverse link is about 2250. Apparently the processor will not be able to process the traffic generated from the air interface. The processor appears to be the performance bottleneck. Improvements are needed to eliminate this problem.

Table 6-5: IP Packet Processing Time versus Packet Size on Forward Link

IP packet size (bytes)	175	303	431	559	687	815	943	1071	1199	1327
Num. of IP fragments	1	1	1	2	2	2	3	3	3	3
Processing time (ms)	0.8	0.8	0.8	1.6	1.6	1.6	2.4	2.4	2.4	2.4

6.2.3 Processor Performance Enhancements

Performance and capacity modeling of the processor is essential in traffic sizing and capacity planning. Another important benefit via performance modeling is to design the enhancements to eliminate any resource bottleneck and improve the overall performance and capacity. Via analyzing the measurement data and looking at the time spent by different tasks, we can identify what are the essential resource consumptions at the processor. For example, for a processor with the VxWorks real-time operating system, the task (i.e., tNettask) that handles the protocol conversion sometimes consumes a large portion of the processing time. In order to reduce the packet processing time, it would be most effective if we can optimize the implementation to reduce the time spent by that particular task.

Another performance enhancement method is to simplify or bypass some processing at the overloaded processor. For example, packet fragmentation usually consumes more time than other processing. We can reduce the PO of the overloaded processor by avoiding the packet fragmentation. This can be done by limiting the size of the packets that pass through the processor. Before entering the overloaded processor, packets with larger sizes can be fragmented by other components, which have larger CPU processing power. This off-loading method balances the load between the components and relieves the overloaded processor.

We use the same example as discussed in Section 6.1.3.2 to illustrate how enhancement can be made to eliminate the performance bottleneck. As discussed earlier, in EV-DO air interface, although the reverse link throughput is less than the forward link throughput, the number of packets per second on reverse link is much higher. The reason is that the packet size on the reverse link is much smaller than the packet size on the forward link. For the specific processor (CRC), the PO is proportional to the rate of the packets rather than the absolute throughput value. Therefore, the reverse link traffic has more stringent requirement in terms of message processing time. In order to reduce the PO, the processing time for the reverse link packet has to be reduced by a large scale.

Detail diagnosis of the processing time consumed by each task is conducted. In this particular example, the tNettask in the VxWorks operating system consumes a large portion of the message processing time. The tNettask mainly does the IP protocol stack processing of routing and fragmentation. In the reverse direction, packets received by the modem board are sent over to the CRC processor through the Firewire bus. Large IP packets will be fragmented at the modem board to fit in the MTU setting of the Firewire bus. The IP packets received by the CRC are already small packets and do not need any fragmentation. The CRC only needs to do simple IP packets routing in the reverse direction. This simple IP protocol processing can be realized without using the full-scale IP stack in VxWorks.

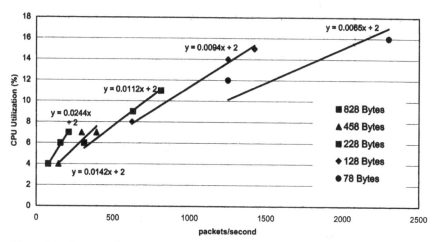

Figure 6-10. Processor Occupancy Measurement on the Reverse Link with the Enhancement

A simplified IP protocol processing via bypassing the VxWorks IP stack is proposed and implemented to improve the reverse link processing performance. Figure 6-10 shows the measurement results with the enhancement of simplified IP processing. As compared to Figure 6-9, it is evident that the processing time is reduced dramatically. The processing time is about 65 μs ~ 112 μs, which is a 7 – 10 times reduction.

With the rate of 2250 packets per second, the PO for the reverse link bearer traffic is

$$P_{RL} = 2250 \times 65 \times 10^{-6} = 14.63\% \tag{6-10}$$

The total PO for the bearer traffic is

$$P = P_{FI} + P_{RL} = 73.7\% \tag{6-11}$$

which is in a normal operating range. The bottleneck at the CRC processor is thus eliminated.

Acknowledgments

The work on the timing adjustment was jointly done with former Bell Lab colleagues Dr. S. Abraham and Dr. A. Sampath.

6.3   Review Exercises

1.   Name the various factors that affect the CPU budget of the main controller card within a UMTS base station.

2.   Discuss the various factors that affect the maximum usable bandwidth of a link between a base station and its associated base station controller.

3.   Write a simple simulation program to repeat the studies done in Section 6.1.2.1 and show that with 768 Kbps, the call carrying capacity is 55.

4.   Using Table 6.3, simulate packet data users and determine the joint capacity with both voice and data users.

5.   Discuss the similarities and differences between designing a UMTS base station and a CDMA2000 1xEV-DO base station.

6.   Describe the various performance enhancements techniques that can be used to improve the processor performance of a CDMA2000 1xEV-DO base station.

6.4   References

[25.101] 3GPP TS25.101, "User Equipment (UE) radio transmission and reception (FDD)", V3.11.0, July 2002.

[25.402] 3GPP TSG RAN, "Synchronization in UTRAN, Stage 2", TS 25.402 V.3.2.0, June 2000

[25.427] 3GPP TSG RAN, "Iur/Iub User Plane Protocol for DCH Data Streams", TS 25.427, V.3.3.0, June 2000.

[Abr02] S. Abraham, A. Sampath, C. Saraydar, M. Chuah, "Effect of Timing Adjust Algorithms on Iub Link Capacity," Proceedings of VTC 2002, pp311–315.

[Chou03] C. Chou, W. Luo, "Comparison of battery consumptions with and without URA_PCH state," Bell Laboratories Internal Memorandum, 2003.

[Chu02] M. Chuah, W. Luo, X. Zhang, "Impacts of Inactivity Timer Values on UMTS System Capacity," Proceedings of WCNC, 2002.

[Sar03a] C. Saraydar, S. Abraham, M. Chuah, A. Sampath, "Impact of Rate Control on the capacity of an Iub link: Single Service Case," Proceedings of ICC 2003.

[Sar03b] C. Saraydar, S. Abraham, M. Chuah, "Impact of Rate Control on the capacity of an Iub Link: Mixed Service Case," Proceedings of WCNC 2003.

Chapter 7

RNC AND RADIO ACCESS NETWORKS DESIGN AND TRAFFIC ENGINEERING

7. INTRODUCTION

As shown in Figure 5.5, a radio access network consists of one or multiple base station controller(s) and tens/hundreds of base stations connected together via the backhaul network. In Chapter 6, we discussed how the base station can be designed. In this chapter, we describe how the base station controller (or Radio Network Controller [RNC] in UMTS terminology) can be designed to meet certain performance requirements. We also elaborate on how the interface bandwidths between the base station controller and base stations can be engineered and discussed techniques that can help to reduce the operation expenses (OPEX) of the radio access networks.

7.1 RNC Design

The main functions of the RNC are the following:
- Manage radio resources. There is an admission control module within the RNC that decides whether a voice call/packet data session will be admitted and the radio resources that need to be allocated.
- Process radio signaling. The RNC processes all the radio signaling coming from the user equipment or from the core network
- Perform call setup and tear down. The RNC needs to exchange signaling messages with the User Equipment (UE) and NodeB to set up radio access bearer service during a voice call or data session establishment process. After the voice call or data session is over, the RNC also needs to exchange signaling messages with the UE and NodeB to tear down the radio access bearer service.
- Perform soft, hard and intersystem handoffs. Based on the radio signal measurements, the RNC has to decide when to direct the

UE to establish soft handoff legs with other NodeBs as well as perform hard handoff or intersystem handoff

- Provide operation, administrative, and maintenance (OA&M) capabilities. The RNC sends keep alive messages across the various interfaces that it uses to communicate with other RNCs, NodeBs, as well as other service access points in the SS7 networks for OA&M purposes.

7.1.1   Overview of Generic RNC Hardware Architecture

The typical hardware architecture of an RNC is shown in Figure 7-1. The intelligence of an RNC, e.g., radio resource management function, resides in the various feature controller cards. These feature controller cards process the NBAP and RANAP signaling messages and make decisions on which voice calls/data sessions to admit. They also house the power control algorithms, and make decisions on when the user should add additional handoff legs etc. For scalability, the feature controller card may itself consists of multiple boards with multiple processors. Such feature controller cards are connected via switched LAN to the various signaling cards, the traffic processing cards, and the gateway cards (referred to as GC). For redundancy purposes, two or more signaling cards and gateway cards are provided in each RNC chassis. Each GC card supports one or more STM-1 ports. The traffic and signaling flows through RNC is shown in Figure 7-2. The signaling traffic will be routed from the gateway card to the respective signaling card and later to the feature controller card. Within the feature controller card, the messages may be passed from one software module (process) to another and/or from one processor board to another. After being processed, new internal or external signaling messages may be sent. The newly generated internal signaling messages are routed from the feature controller card to the signaling card, while the external signaling messages are routed to either the Core Network or NodeBs via the gateway card.

7.1.2   RNC Capacity

Before discussing some design guidelines for an RNC, we first describe several important capacity metrics that are typically reported in the datasheet of any RNC vendor. Normally, an RNC is designed to handle a certain capacity, e.g., x Erlangs of voice traffic (with 12.2 Kbps per channel). Assuming an average channel holding time of y s, then the maximum number of busy hour call attempts (BHCA) is $x \cdot 3600/y$. For an example, if x = 4000 Erlangs, y = 100 s, then the maximum BHCA an RNC can handle is 144K BHCAs. Each RNC can support a certain maximum number of cells

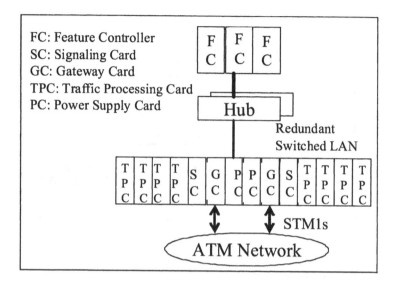

Figure 7-1. Generic Hardware Architecture for RNC

and NodeBs (typically we assume 3 cells/NodeB). It has a certain number of STM-1 interfaces, and a certain maximum raw throughput over the Iu, Iur, Iub interfaces. Typically, the maximum raw Iub throughput is 150% of the maximum raw Iu throughput while the maximum raw Iur throughput is designed to be about 1/3 of the maximum Iu throughput. Sometimes, the data sheet of an RNC will also indicate the maximum number of cell-connected data users and UTRAN Registrations Area (URA) connected users it can handle simultaneously.

Such capacity goals are then used to design each RNC component, e.g., the feature controller card, the gateway & signaling cards, and the traffic processing cards. Based on the desirable capacity design goals on the number of simultaneous voice and data calls that need to be supported, one can determine the amount of signaling and bearer traffic that will be generated using appropriate traffic and mobility models.

In subsequent sections, we first give an example of the circuit/packet switched traffic model. Then, we describe how these models are used to determine the amount of signaling/bearer traffic that will be generated per RNC. The amount of signaling traffic generated also depends on other factors, e.g., inactivity timer of Radio Access Bearer, etc.

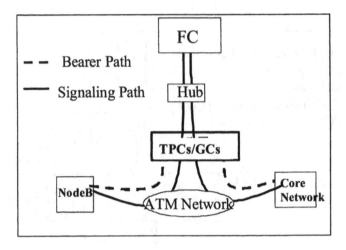

Figure 7-2 Bearer/Signaling Flow Through RNC

7.1.3   Traffic Model Revisited

In Chapter 3, we described how one can derive appropriate traffic models for different applications. Here, we give a summary of the traffic models typically used for circuit switched and packet switched services.

7.1.3.1 Circuit Switched Traffic Model

Table 7-1 shows the traffic parameters used to describe the circuit switched services such as voice and videophone. To come up with a good traffic model, one needs to make assumptions based on past experience on the busy hour call attempt (BHCA) per subscriber for each service type. One also needs to make assumptions on the percentage of call originations/terminations, the number of soft handoffs per call, the number of location updates per call, the bit rate generated per voice user or per videophone user, the probability of page failure upon the first and second page. Using this traffic model, one can compute the signaling and bearer traffic generated by the active subscribers that produce the maximum number of BHCAs that the RNC is designed for.

There are many circuit switched signaling procedures. One can determine the 10 most common procedures for a voice/videophone scenario, e.g., RRC connection establishment/release, radio access bearer establishment/release, radio link addition/deletion, NAS message transfer, Core Network (CN) paging, UE dedicated measurement request, and NodeB dedicated measurement request. Then, one identifies how many of the above 10 procedures will be activated in each usage scenario, e.g., UE terminated call, UE originated call, location update. Using the parameters in the traffic models and the mobility

Table 7-1 Circuit Switched Traffic Model

General Input Parameters	
Total subscribers supported by an RNC	240,000
Circuit Switched Service	
Circuit switched call attempt weighting factor	100%
Voice BHCA per subscriber	0.6
Videophone BHCA per subscriber	0.2
Videophone usage weighting factor (average per subscriber)	0%
Total number of voice calls per BH in the RNC	1441,000
Total number of videophone calls per BH in the RNC	0
Total Erlang due to circuit switched calls	4000
Total number of SMS going thru Iu-cs per BH	0
Circuit Switched Traffic Characteristics	
Maximum voice BHCA per MSC	300,000
Average call holding time (ACHT) per voice BHCA	100
Average call holding time (ACHT) per videophone BHCA	60
Voice call termination rate	70%
Voice call origination rate	30%
Voice call voice activity factor (includes SID updates)	50%
SMS going through MSC	144,000
SMS termination rate	70%
SMS origination rate	30%
SHO per BHCA	5
Location updates per BHCA	1.5
Location update holding time	3
Voice bit rate	12.2
Videophone average bit rate	48.0

Table 7-2 Common Procedures (Voice Scenarios) in RNC and Their Message Frequencies

	Frequency (/sec)	RRC Connection Establishment	RRC Connection Release	Radio Access Bearer Establishment	Radio Access Bearer Release	Radio Link Addition	Radio Link Deletion	NAS Message Transfer	UE Dedicated Measurement Req
Use Case									
MO call	28	1	1	1	1	5	5	16	25
MT call	12	1	1	1	1	5	5	16	25
Location Update	60	1	1					2	0.75
Inter-Process (MO/MT)	2	3	0.6	4.5	4.4	4.4	1	0	
Inter-Process(LU)	1	4	0	0	0	0	1	0	
Inter-Processor (MO/MT)	1	2	0.25	2.4	2.4	2.4	0	0	
Inter-Processor (LU)	0	2	0	0	0	0	0	0	
TPC-in(MO/MT)	3	3	2.5	2.5	2	2	0.5	1	
TPC-in(LU)	4	2	0	0	2	2	0.6	1	
TPC-out(MO/MT)	3	3	5	4	2	2	0.5	0	
TPC-out(LU)	2	2	0	0	2	2	0.6	0	
ALCAP(MO/MT)	2	2	1	1	1	1	0	0	
ALCAP(LU)	2	2	0	0	1	1	0	0	
Use Case Freq									
MO call		28	28	28	28	140	140	448	700
MT call		12	12	12	12	60	60	192	300
Location Update		60	60	0	0	0	0	120	45
Total		100	100	40	40	200	200	760	1045
Inter-Process		140	360	24	180	880	880	760	0
Inter-Processor		40	200	10	96	480	480	0	0
TPC-in		360	240	100	100	400	400	392	1045
TPC-out		240	240	200	160	400	400	392	0
ALCAP		200	200	40	40	200	200	0	0

model, one can determine the number of activated usage scenarios per second. Using this number, one can determine the frequency of the invocation of each signaling procedure as shown in Table 7-2. Each procedure generates a certain number of NBAP & RANAP messages and requires a certain amount of CPU and memory resources.

Let us go through some numbers shown in Table 7-2 to see how they are derived. From the voice traffic model in Table 7-1, one can derive that the total Busy Hour Call Attempt (BHCA) is the product of the number of voice subscribers and the average call attempt per busy hour from each subscriber, i.e., $240000 \cdot 0.6 = 144,000$. With a 70%/30% split of Mobile Originated/Terminated calls, one can determine that the MO and MT calls/s is 28 and 12, respectively. Similarly, since we assume 1.5 location updates per call, the total location update per second is 60. In Table 7-2, we tabulate the number of inter-process, inter-processors messages, the number of messages that traverse in/out of the Traffic Processing Card (TPC), and the number of ALCAP messages that need to be invoked per major procedure for each scenario. Note that these numbers are given as an example. Different RNC designs give different numbers for TPC-in/TPC-out, Inter-Process and Inter-Processor messages. With these numbers, one can derive the total number of Inter-Process, Inter-Processor, TPC-in, TPC-out, ALCAP messages for the major procedures.

Consider the example of Radio Resource Control connection establishment procedure. From Table 7-2, we see that the number of TPC-in messages generated per MO/MT call during the RRC connection establishment procedure is 3 while the number of TPC-in messages generated by Location Update (LU) is 4. So, the total TPC-in messages caused by both MO/MT calls and LUs is $(18 + 12) \cdot 3 + 60 \cdot 4 = 360$ which the number shown in second column, second row in Table 7-3. Similarly, to compute the total TPC-out messages generated, we merely add those generated by the MO/MT calls and those generated by the LU events which gives $(18 + 12) \cdot 4 + 60 \cdot 2 = 240$. By cranking through all the procedures and scenarios, one can fill up all the columns and rows as shown in Table 7-3.

Table 7-3 Summarized Signaling Message Frequencies for Different Procedures

Msg Freq per use case	TPC-in	TPC-out	SS7-in	SS7-out	ALCAP	InterProcess	InterProcessor	Total
RRC connection establishment	360	240	0	0	200	140	40	980
RRC connection release	240	240	0	0	200	360	200	1240
Radio access bearer establ.	100	200	40	40	40	24	10	454
Radio access bearer release	100	160	40	40	40	176	96	652
Radio link addition	400	400	0	0	200	500	300	1800
Radio link deletion	400	400	0	0	200	500	300	1800
NAS msg transfer	360	360	360	360	0	0	0	1440
UE dedicated measurement	1045	0	0	0	0	0	0	1045
Cell common measurement	96	0	0	0	0	0	0	96
Total	3101	2000	440	440	880	1700	946	9507

7.1.3.2 Packet Switched Traffic Model

Our packet switched traffic model uses the "Always-On Approach" for packet data users. The "Always-on Approach" means once the user activates a data session, it will toggle between URA-PCH and cell-connected mode and will not return to idle mode. The user will enter idle mode only if he/she powers down the wireless device. Five possible applications are considered: WAP, SMS, Email, Web browsing, and audio streaming. The traffic model for each application is as shown in Table 7-4. We assume the application mixture is as follows: 28% WAP, 50% SMS, 10% email, 10% Web browsing, and 2% audio streaming. With this traffic mixture distribution, one can compute the average number of data transactions per session to be $(0.1 \cdot 18 + 0.1 \cdot 23 + 0.02 \cdot 2 + 0.28 \cdot 20 + 0.5 \cdot 1) = 10.24$. One can also determine the average number of downlink/uplink packets per transaction, the average downlink/uplink packet size, the average number of reactivations per session, the average session length etc as shown in Table 7-4. Table 7-5 shows the assumptions for some additional traffic descriptors for the packet switched traffic model, e.g., the registration area (RA) location updates and the UTRAN registration area (URA) location updates per attached data user during busy hour. Some parameters listed in Table 7-5 are computed from the numbers shown in Table 7-4 e.g. given that the average number of transactions per session with the given traffic mixture is 10.24, one can derive the total number of data transactions during BH as $85680 \cdot 1.0 \cdot 10.24 = 877,363$.

Another example is the number of cell-connected data users that can be supported per RNC. Assume that there are 48 NodeBs attached to each RNC, each NodeB has 3 cells, and that each cell can support 21.8 64 Kbps cell-connected data users (such number can be determined using the methodology described in Chapter 5). Then, the total number of cell-connected users that can be supported per RNC is $21.8 \cdot 48 \cdot 3 = 3150$.

As in the circuit voice case, we also need to determine the total number of signaling messages generated by packet data users. We can use the same methodology as in the circuit voice case. First, we identify the most common packet switched procedures. The 16 most common packet switched procedures are Radio Access Bearer Establishment/Release, Radio Link Addition/Deletion, NAS Message Transfer, Cell_DCH->URA_PCH state transition and vice versa, Cell_FACH->Cell_DCH state transition and vice versa, URA location Update, Cell Update at SRNC, Cell Update at DRNC, Core Network (CN) Paging, UTRAN paging, UE/NodeB dedicated measurement, Cell common Measurement. Each procedure generates a certain number of signaling messages, and requires a certain amount of CPU resources as well as memory storage.

Table 7-4 Summarized Packet Switched Traffic Models for Different Applications

Name of the application	E-mail	Web	VoIP	Audio	WAP	SMS
% of session usage	10%	10%	0.%	2%	28%	50%
Number of transaction per session	18	23	240	2	20	1
Data Call Termination rate	56%	0.%	45%	0.%	0.%	70%
No of Uplink pkts per data session	104	920	4800	600	40	4
No of Dnlink pkts per data session	40	920	4800	1600	40	4
Tot. Uplink Kbytes per data session	42	81	288	48	2.5	0.1
Tot. Dnlink Kbytes per data session	56	757	288	576	31	0.7
Per data transaction size uplink	4.2	3.5	1.2	24	0.1	0.1
Per data transaction size dnlink	5.6	32.9	1.2	288	1.6	0.7
User Think Time (sec)	127	26	0.6	40	27	0
Avg "On+Off" time per trans (sec)	130	33	1.2	216.4	29	2
Avg uplink pkt size (bytes)	400	88	60	80	62	13
Avg dnlink pkt size (bytes)	1400	822	60	360	775	172
% of transactions in traffic mix	11	24	0	0	59	5
% of uplink pkts in trffic mix	8	72	0	9	9	2
% of dnlink pkts in traffic mix	3	65	0	23	8	1

Table 7-5 Packet Switched Traffic Model

Packet Switched Service	
Packet switched data session attempt weighting factor	100%
BHDSA per Attached Data User	1.0
% subscribers with packet data service	60%
% data subscribers attached during BH	60%
Total data subscribers supported in RNC	144,000
total data subscribers attached via RNC during BH	85,680
Packet Data Traffic Characteristics	
Total Number of Data Transactions per BH per RNC	877,363
RA Location Updates per Attached Data Users per BH	0.2
URA Location Updates per Attached Data Users per BH	1.5
Max Simult. Cell-Connected Users	10,000
Average Simultaneous Cell-Connected data users during BH	3150
Total Number of Reactivations per BH per RNC	695,603

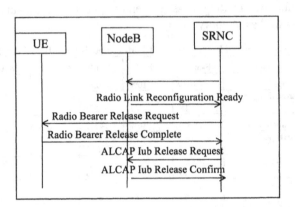

Figure 7-3 Signaling Messages Involved in the Cell-DCH to URA-PCH Transition

Let us consider the Cell-DCH->URA_PCH transition in detail to see what messages need to be sent. Typically, a user is moved from the Cell-DCH to the URA-PCH state due to RRC inactivity timer expiration. The signaling messages involved in this transition are shown in Figure 7-3. The SRNC will send a NBAP radio link reconfiguration prepare message to the NodeB for deleting the DCH. The NodeB will respond with a radio link reconfiguration ready message. This is followed by a radio link reconfiguration commit from the SRNC. The SRNC then sends a radio bearer release message to the User Equipment (UE). The UE later sends a radio bearer release complete message to the SRNC. Then, the SRNC and NodeB exchange ALCAP Iub Release Request/Confirm message.

When the UE is in the URA_PCH state and a downlink packet destined to the UE arrives at the SRNC, then the SRNC will send messages to transition the UE from the URA_PCH state to the Cell_DCH state. The signaling messages involved in this transition are shown in Figure 7-4. The SRNC sends a paging message to the UE. The UE then responds with a Cell Update message. Then, the SRNC sends a radio link setup request message to the NodeB. The NodeB responds with a radio link setup response message. The SRNC and the NodeB also exchange the ALCAP Iub establish request/confirm messages. In addition, the SNRC and the NodeB exchange dowlink/uplink synchronization messages. To complete the procedure, the UE will send a radio bearer reconfiguration complete message to the SRNC. One can compute the frequency at which a certain signaling procedure will be invoked assuming a certain application mixture and the traffic models for each application. One can also count the number of signaling messages that need to be exchanged between the SRNC and the NodeB, and between the SRNC and the UE for each procedure. In addition, based on the designed RNC software and hardware architecture, there may be some internal messages that need to be exchanged between different cards and different modules within the same card.

For always-on packet data users, the procedures that have the highest frequencies are those that transition the data users from the Cell_DCH to the URA_PCH states, and from the URA_PCH to the Cell_DCH states. Using the traffic models of different applications as tabulated in Table 7-4, one can come up with the average on time, the average off time, and the average session time of the 64-Kbps data users. In addition, one can determine the achievable multiplexing gain for 64-Kbps data users. The achievable multiplexing gain (MG) is computed as follows:

$$Multiplexing\,Gain = \frac{On\_time(64/144/384) + Off\_time}{On\_time(64/144/384) + Inact\_timer}$$

After determining the MG, one can compute the number of 64-Kbps users per cell per hour, Y, to be

$Y = MG(64)*$Number of 64-Kbps DCH users (accounted for softhandoff)$*3600/($Tot_BHDSA$*$Avg_Session_Dur(64))

where MG(64) is the multiplexing gain for 64-Kbps data users, Tot_BHDSA is the total busy hour data session attempt and Avg_Session_Dur(64) is the average session duration for 64-Kbps data users (average over the application mixture). Using the same example where we can support 21.8 64Kbps DCH users, and 1.5 softhandoff/softer handoffs/user number, Y can be computed to be about 235 (5.3*21.8/1.5*3600/(1*1197.4)) assuming Total_BHDSA is 1/hr and Avg_Session_Dur(64) = 1197.4 s.

The number of packet calls per cell per hour for 64-Kbps data users, Z, can be derived as follows:

$Z =$ Number of DCH users (accounted for softhandoffs) *3600
    / (On_time_64K +Inact_timer)

where On_time_64K is the average on time for 64-Kbps data users (average over different applications) and Inact_timer is the value of the inactivity timer (typically set to 5 s). Using the same example, Z can be computed to

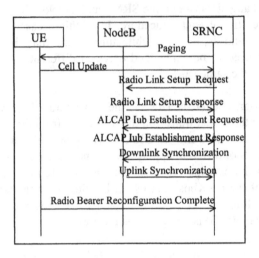

Figure 7-4 Signaling Messages for URA-PCH to Cell-DCH Transition

be 14.6*3600/(12.5+5) = 3004. This number can be considered as the total number of reactivations per cell per hour. For an RNC that supports 48 NodeBs where each NodeB supports 3 cells, one can easily determine that the total reactivations per RNC per second is 3004*48*3/3600 = 120/s.

Using the total number of reactivations per RNC per second number, we can derive the number of messages/s generated by the radio link addition/deletion procedures (rows 6 and 7 in Table 7-6). Each of these procedures involves two ALCAP messages, two TPC-in, two TPC-out messages. Thus, one see that the appropriate columns have numbers close to (120*1.3*2) = 312/s. Two interprocessor communications are invoked per reactivations of a user (meaning reactivations due to softhandoffs do not incur interprocessor communications) so the total rate of interprocessor communications caused by the radio link addition activity is close to 120*2 = 240/s.

With such estimations, one can construct a table similar to Table 7-6 and determine how much bandwidth, CPU, memory are required to support the number of simultaneous active data users in each RNC. Then, one can choose the link interface (whether it is E3 or STM-1), the number of such interfaces that need to be supported per RNC, the CPU and its associated clock rate chosen for each card and how many such cards are required to support the desirable capacity goals.

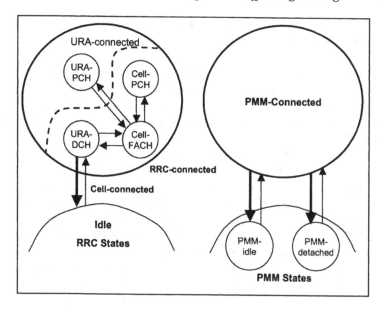

Figure 7-5 Packet Data State

Table 7-6 Message Frequencies for the 16 Most Common Procedures for Packet Data Users

Msg Freq per use case	TPC-in	TPC-out	SS7-in	SS7-out	ALCAP	InterProcess	InterProcess	Total
RRC connection establishment	11.7	8.3	0	0	3.3	4.7	1.4	29.4
RRC connection release	8.3	8.3	0	0	3.3	12	7.2	39.1
Radio Access Bearer Establ.	8.3	16.7	3.3	3.3	3.3	1.8	0.8	37.5
Radio Access Bearer Release	8.3	13.3	3.3	3.3	3.3	18	8	57.5
Add Initial Radio Links	348	510	0	0	232	32.5	4.6	1127.1
Radio Link Addition	315	315	0	0	158	378	227	1393
Radio Link Deletion	315	315	0	0	158	378	227	1393
NAS Msg Transfer	214	214	214	214	0	0	0	856
Cell_DCH-> URA_PCH	290	290	0	0	0	870	435	1885
Cell_FACH-> Cell_DCH	0	0	0	0	0	0	58	58
Cell Update at SRNC	460	460	0	0	0	460	0	1380
Cell Update at DRNC	0	0	0	0	0	0	0	0
URA Location Update	36	36	0	0	0	72	0	144
UTRAN Paging	0	44	130	0	0	18	50	242
UE Dedicated Measurement	790	0	0	0	0	0	0	790
Cell Common Measurement	96	0	0	0	0	96	0	192
Total	2900.6	2230.6	350.6	220.6	561.2	2341	1019	9623.6

7.1.4   Impacts of RAB Inactivity Timer Value on Signaling Traffic and Power Consumption

Figure 7-5 shows the various states that a packet data user may be at. When the UE is switched on, it registers with the network but has no dedicated connections assigned to it. The UE is then said to be in idle state. Once a UE activates a call or a session, and is given a dedicated connection, then the UE will be in Cell-DCH state. For a packet data session, there may be idle periods in between bursts of data traffic. When such idle periods exceed the inactivity timer value, the UE will be moved out of the Cell-DCH state to a URA-PCH state where the UE is still connected but it has no dedicated channels assigned to it. A UE in this state needs to be paged over a UTRAN registration area (URA). Sometimes before switching a UE to the URA-PCH state, the UE may be moved to the Cell-FACH state where the UE has no dedicated air interface resources but communications can take place via common channels. To move the UE from the idle state directly to the Cell-DCH state requires about 42 messages while to move the Cell-DCH state to the idle state requires 30 messages. Considering the message transmission delay, physical layer interleaving, coding, and guard timer values, etc, the total delay for a transition from idle to Cell-DCH can be up to 5 s. However, to move from the URA-PCH state to the Cell-DCH state requires only 12 messages and to move from the Cell-DCH state to the URA-PCH state requires about 8 messages. The transition from URA-PCH to Cell-DCH requires only 1 s without paging. Thus, it seems that by choosing appropriate value for the inactivity timer, one may be able to release radio resources from packet data users that are idle and at the same time be able to switch such idle users fast into the Cell-DCH state whenever new data for the users arrive. It is challenging to pick a suitable value for the inactivity timer. Choosing a high value reduces the ability to share the radio resources. However, choosing a small value may increase the signaling load for the user may be switched too frequently into the URA-PCH state and later switched back to the Cell-DCH state.

In order to understand whether the allocated codes are used efficiently, one can define an efficiency metric, E [Chou03], as follows:

$$E = \frac{t_{on}}{(t_{on} + t_D + t_{inact})}$$

where t_{on} is the average on time of a data transaction, t_D is the average network setup time, and t_{inact} is the average inactivity time. The average inactivity time is defined as the time when radio resources are held during the inactivity period after a data transaction has completed until either the arrival of the next data transaction or the releasing of radio resources upon timer expiration.. t_D can be computed using the following equation:

$$t_D = D \times \Pr(t > T) = D \times (t - F(T))$$

where $F(t)$ is the cumulative distribution function of the user think time and D is the reactivation time for different approaches. D is 1 s for the URA_PCH approach and 5 s for the idle state approach. t_{inact} can be computed using the following equation where $f(t)$ is the probability density function of the user thinking time and T is the inactivity timer value.

$$t_{inact} = T \times (1 - F(T)) + \int_0^T tf(t)dt$$

With an average web page size of about 60 Kbytes, an average user think time of 30 s, and assuming exponential distribution for user think time, 64 Kbps downlink/uplink data rate, and an inactivity timer value of 5, one can easily compute t_D to be 0.85 s, t_{ON} = 9 s, t_{inact} = 4.6 s. So, E can be computed to be about 62% for Solution 1 (refers to Figure 7-6) where a packet data user will be transitioned from Cell_DCH state to URA_PCH state after being idled for a certain period, T. If we want Solution 2 (where a packet data user will enter into idle period after being idle for T)

to have the same t_D, then the inactivity time needs to set to about 52 s. With $T = 52$ s, we can calculate t_{inact} and E. E is only about 26% for Solution 2. Thus, we see that Solution 1, where the data users are transitioned into the URA_PCH state after they are idle for some time, achieves a higher efficiency metric than Solution 2, in which the data users are transitioned into the idle state after they have been idle for a period of time.

Another benefit of the URA-PCH state is in terms of the savings in the battery life of a UE. Let us use a simple example to illustrate this energy saving. Assume that a UE in Cell-DCH is transmitting 64 Kbps data compared with a UE in URA-PCH having a paging channel at 128 Kbps with a paging indication channel cycle of 1/256. A rough comparison of the battery power consumption during that period can be given by

$$\text{Power Consumption Ratio} = \text{Activity Ratio} \times \frac{\text{Paging Channel BW}}{\text{Data Channel BW}}$$

$$= \frac{1}{256} \times \frac{128}{64} = \frac{1}{128}$$

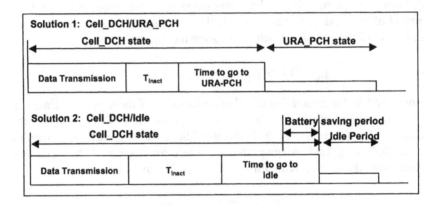

Figure 7-6 Comparison of Battery Consumption for Both Solutions

The period whereby the power consumption ratio holds may be as long as 4 s (refers to Figure 7-6) considering the message transmission delay in Solution 1.

7.1.5   Radio Resource Management

In UMTS, transmissions from different sources are separated by channelisation codes. The channelization codes are based on the Orthogonal Variable Spreading Factor (OVSF) technique which was originally proposed in [xx]. The use of OVSF codes allows the spreading factor (and hence the user data rate) to be changed while maintaining the orthogonality between different spreading codes of different lengths. The codes are picked from the code tree as shown in Figure 7-7. There are some restrictions as to which of the channelization codes can be used by a single source. A physical channel may use a certain code in the tree only if no other physical channel using the same code tree is on the underlying branch, i.e., using a higher spreading factor code generated from the spreading code that is intended to be used. For example, if a physical channel already uses code $c_{2,0}$, then no other physical channels can use the code $c_{1,0}$ or anything below the code tree $c_{2,0}$ e.g. $c_{4,0}$. The radio resource management module within the RNC runs the code assignment scheme that allocates orthogonal codes to data users. Since the number of available orthogonal codes is limited if the secondary scrambling code is not used, designing an efficient code assignment scheme is highly important. An efficiently designed scheme allows more data users to be admitted into the system and hence provides higher multiplexing gain.

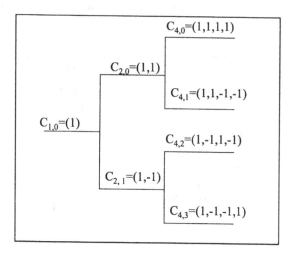

Figure 7-7 Channelization Code Tree

Typically, some small number of codes is reserved for common packet channels, e.g., one or two 64-Kbps forward access channel (FACH). The rest will be allocated as dedicated traffic channels to the users.

As shown in Figure 7-8, packet data service users must be in the Cell_DCH or Cell_FACH state to transmit data [25.331]. In the Cell-DCH state, the user is provided with a dedicated channel with a fixed data rate. In the Cell_FACH state, the user shares the forward access channel (FACH) with other users so the bandwidth available to a user in the Cell_FACH state is much lower than that in the Cell_DCH state. The number of users that can be kept in the Cell-DCH state depends on the data rate assigned to each user. Assuming that all users are assigned a dedicated data rate of 384 Kbps each, then only a maximum of seven such users can be simultaneously kept in the Cell_DCH state if the secondary scrambling code is not implemented.

Packet data users are known to generate bursty data traffic. For example, users running applications such as email and Web browsing incur certain periods of inactivity when they read the emails or the downloaded Web pages. In addition, such applications have a more relaxed delay requirement. A typical person can tolerate a few seconds (1 -3 s) delay for a Web page or email download and hence a few hundreds milliseconds of delay in obtaining a DCH may be tolerable. These two properties of relaxed delay requirement and the bursty nature of the traffic stream can be exploited to admit more data users so that the limited number of codes can be shared more efficiently. One way of doing this is to have the users release the codes

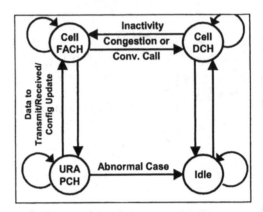

Figure 7-8 Transition State of Packet Data Users

assigned to them when they are idle and request for new codes when new data arrives. If the switching can take place such that a user receives a DCH when the user is ready for the next burst of data, then the user's perceived response time will not be noticeably degraded. The released codes can be used to serve other users. Assigning DCHs dynamically on demand in this manner would allow more data users to be admitted into the UMTS network while still guaranteeing a reasonable quality of service to all of them.

In this section, we describe four algorithms for the dynamic allocations for DCHs among packet data users in a UMTS network.[11] The four algorithms are:

1.   Fixed inactivity timer (FIT) algorithm

In this algorithm, when a user's transmit buffer level drops to zero, the inactivity timer is started. The timer is reset to zero if a new packet is received at the RNC for the user or new uplink packets are received. When the timer exceeds a certain preset threshold, the procedures for releasing the DCH are initiated. The user is transitioned to the Cell_FACH state. The freed DCH can be allocated to other users.

Choosing a good value for the fixed inactivity timer is not easy. A high value will induce fewer switches between the Cell_DCH and

---

[11] This work was done by Dr. S. Abraham, Dr. M. Chuah when they were at Bell Laboratories and documented in an unpublished memorandum.

Cell_FACH state and hence require smaller signaling bandwidth and processing capacity. However, a high value causes poor sharing of the DCHs. A low value can cause unnecessary switches between the Cell_DCH and Cell_FACH state and creates a drain on the system processing capacity and UE power. However, the utilization of the DCH might be better with a lower inactivity timer threshold.

2. Least recently used (LRU) algorithm

In this algorithm, a user is allowed to remain in the Cell_DCH state in the absence of uplink/downlink transmission, as long as there are no other users requesting a DCH. When another user requires DCH bandwidth, the inactivity timer of all users in Cell_DCH state is scanned. The user that has the highest inactivity timer value will be switched to the Cell_FACH state. The freed DCH is allocated to the user that requests for it.

3. Combined FIT and LRU algorithm

The FIT and LRU algorithm both require the use of an inactivity timer. The two approaches can be combined to overcome some of the potential problems associated with either approach. The combined algorithm works as follows: when the inactivity timer has exceeded a certain preset threshold, the user is moved to the Cell_FACH state. The LRU procedure is used to choose a user whose resources will be released if there is no user whose inactivity timer has expired.

4. Adaptive algorithm

In this algorithm, the total queue size (including new arrivals) is monitored and a lowpass-filtered version of this variable is used to decide when to switch a user to either the Cell_DCH or the Cell_FACH state. A switching decision is made every T s. To avoid a ping-pong effect, a session must remain in a given state for w s before it can be switched to another state. The switching decision is explained in the following paragraphs:

A variable x is used to track the sum of new uplink/downlink RLC PDUs arriving within the update interval as well as the RLC queue size (in terms of RLC PDUs). A variable y is defined to be a filtered version of x where the filtering constant is a.

$$y_n = a\, y_{n-1} + (1-a)\, x_n$$

When this variable y is less than a threshold α and the user is in the Cell_DCH, then the user is switched to the Cell_FACH state. If the variable y is more than another threshold, β, and the user is in the Cell_FACH state, then the user is switched to the Cell_DCH state.

If $(y < \alpha$) && user-state = Cell_DCH
    Switch to Cell_FACH
If $(y > \beta$) && user-state = Cell_FACH && DCH is available
    Switch to Cell_DCH.

Some simulation studies are performed to compare these four algorithms using NS simulator [NS], and the Web page model described in [Mah97]. In this web page model, the average number of objects within a page is 5 and the average object size has a mean of 7 Kbytes. The think time is assumed to be exponentially distributed with a mean of 25 s and the number of pages downloaded is geometrically distributed with a mean of 23. In the simulation, they assume that the power allocated to a user will be updated at regular intervals. The required power allocation is based on an empirical joint distribution of Ior (power received by a user from its own sector) and Ioc (power received by the user from other sectors) that was obtained from field tests. The power allocation is based on the Ior/Ioc ratio and the data rate of the user. If there is enough power, the user will be allocated 384 Kbps. Otherwise, lower data rates such as 128 Kbps or 64 Kbps or 0 Kbps will be allocated. Once a user is allocated the power, we assume that the user consumes its full power allocation during the TTIs where there is data to be sent and 10% of the allocated power in TTIs where no data is sent.

For the results tabulated in Table 7-7, the following assumptions are made:
- The time required to switch from one state to another is 0.5 s.
- The one way delay from the server to the UE is assumed to be around 40 -50 ms.
- For FIT, the fixed timer threshold is set at 3.5 s.
- For LRU, to prevent excessive switching for instances when there is a short duration with an empty buffer, we ensure in our simulation that a buffer needs to be empty for at least 100 ms before its user will be considered.
- For Adaptive Algorithm, a user must remain in a state for $w = 1$ s before that user can be switched to another state. β is chosen such that 75% of the object size is larger than this value. For web traffic, β is chosen to be 10 PDUs (400 bytes). With an update interval of t s, the adaptive filter emulates an RC filter with time

constant *at/(1-a)*. So, a is chosen such that *5at/(1-a)* is equal to 0.5 s. α is chosen such that the bandwidth required to drain the backlog α within the update time *t* is less than $1/10^{th}$ of the FACH bandwidth, i.e., $\alpha \leq ft/3200$ (recall that α is expressed as multiples of 40 bytes RLC PDU). For the reported simulation results, two update intervals are used: namely 1 TTI and 5 TTIs. When the update interval is one TTI, the values of *a, α, β* are set at 0.84, 0.2, and 10 PDUs. With 5 TTIs, the values of *a, α, β* are set at 0.5, 1.0, and 10 PDUs. The FACH data rate used was 32 Kbps.

Table 7-7 shows the results with a session arrival rate of 1 every 40 s (a low load case). The results indicate that FIT and adaptive algorithms has a mean DCH holding time of no more than 7 s. Other statistics collected are tabulated in Table 7-8. The page download time for FIT and FIT+LRU algorithms is very similar. This is expected since at low load, no queueing delay is incurred in obtaining a DCH. The page download time for the adaptive algorithms is slightly higher because the filtered variable needs some time to reach the switching threshold. However, the power used for the LRU algorithm is very high while the number of switches between the FACH and DCH states is least with the LRU algorithm. The power utilization improves with the FIT+LRU approach. Among the 4 algorithms, the adaptive algorithm with an update period of 5 TTIs and the FIT+LRU approach seem promising. With a higher session arrival rate, the simulation results show that the FIT algorithm can no longer sustain the high load without building up a long queue of DCH requests. TCP sessions may fail to operate at high load with such a long delay in obtaining DCH service using the FIT algorithm.

Table 7-7 DCH Holding Time with Different DCH Allocation Schemes

Algorithm	DCH Holding Time	Time to Release DCH	Total time Buffer > 0 During Download Time	Total time Buffer = 0 During Download Time
Fixed Inactivity Timer (3.5 s)	7	3.51	1.22	2.27
LRU	19.27	10.21	1.73	7.33
Adaptive (1 TTI)	3.92	1.02	1.12	1.78
Adaptive (5 TTI)	4	1.09	1.12	1.79
Fixed + LRU	6.92	3.44	1.21	2.26

Only Web users are considered for the above results. Simulation studies are also performed using other traffic models. Interested readers can refer to the third question in the review exercise section for a simplified analytical model that can be used to evaluate the performance of the code allocation scheme for the low load case.

Table 7-8 Other Useful Statistics: Average Power Usage, Average No of State Switching/Page, Average Throughput with Different DCH Allocation Schemes

Parameter	FIT (3.5s)	LRU	FIT + LRU	Adap: 5TTI	Adap: 1TTI
Occupancy of DCH	8.66%	8.76%	8.62%	8.36%	8.51%
Allocated bandwidth in DCH state	383.68	383.74	383.87	383.91	383.98
Power used (dB)	-21.10	-19.89	-21.23	-21.46	-21.35
Av. no. of switches per page	1.65	1.21		1.77	1.76
Mean page download time	3.87	3.73	3.73	4.16	4.25
Number of pages sent per completed user	23.02	22.21	22.5	21.77	22.35
Number of users completely served	981.00	980.00	983	979.00	977.00
Mean number of sojourning users	15.65	15.12	16.32	14.91	15.38
Mean page size		36.01 KB	35.96 KB	36.10 KB	36.16 KB	35.86 KB

With a higher session arrival rate (one arrival every 20 s), the simulation results show that the FIT algorithm no longer can sustain the high load without building up a long queue of DCH requests. TCP sessions may fail to operate at high load with such a long delay in obtaining DCH service using the FIT algorithm.

7.1.5.1 Impacts of Variable Backhaul Delay on the Designed Algorithms

The above results assume that the one-way delay between the Web server and the mobile station is a fixed delay of about 50 ms. In real networks, the delay between the Web server to the GGSN may be a variable. Thus, another simulation experiment was carried out assuming an additional exponentially distributed backhaul delay of either 100 ms or 200 ms. In this second set of experiment, we assume that 39% of the users run HTTP applications, 48% of the users run email, and 13% of the users run FTP programs. The results tabulated in the second column of Table 7-10[12] indicate that the earlier suggested method for selecting a, α, and β does not work well when a long backhaul delay is considered. Here, we propose another method for choosing appropriate values for a, α, and β. Our goals in setting the parameters are such that (i) the user state can be switched from the Cell FACH to the Cell DCH state quickly when the download of a new page begins, (ii) transitions can be prevented due to the jitter on the packet

---

[12] Subsequent work was carried out with by Dr. Abraham, Dr. Chuah with Dr. D. Calin from Bell Laboratories and documented in another internal memorandum.

interarrival times, (iii) the user state can be transitioned from Cell DCH to Cell FCH state when the data utilization is low. To prevent transitions taking

Table 7-9 Parameter Settings for the 100 ms Core Network Delay Case

Update Interval (T)	$T/\ln a$	D	a	β	α
100 ms	500 ms	500 ms	0.8187	6.3	1
200 ms	600 ms	500 ms	0.7165	6.3	2
300 ms	600 ms	500 ms	0.6065	6.3	3

Table 7-10 Simulation Results with LRU and Adaptive Algorithms with Old/New Parameters

Parameter	LRU	Adap: Old Param	5 TTI	10 TTI
Occupancy of DCH	4.63%	3.42%	4.27%	5.16%
Mean total power allocated (dB)	-20.91	-23.38	-22.89	-22.16
HTTP: Number of switches per page	0.48	5.20	1.81	1.70
Email: Number of switches per page	0.86	3.06	3.24	1.68
HTTP: Mean page size	61.72	60.24	57.54	57.42
Email: Mean page size	41.37	41.29	41.04	40.92
HTTP: Mean user perceived bandwidth	25.84	19.83	24.51	24.42
Email: Mean user perceived bandwidth	65.0	45.26	55.90	61.82
FTP: Mean user perceived bandwidth	124.11	106.87	122.16	124.11

place as a result of the jitters on the packet interarrival times, a value D is picked such that $\Pr\{rtt > D\} \le 0.05$ and a is chosen such that $\dfrac{T}{\ln a} = T\left[\dfrac{D}{T}\right]$,

where T is the update period. Recall that $T/\ln a$ is the time constant for the algorithm to react. We pick a value k such that 75% of the object size will be greater than k. Then, we pick β such that the y_n value is β after n updates with no new arrivals or departures with $y_0 = k$. α is still chosen such that the bandwidth required to drain a backlog of α within an update interval is less than $1/10^{th}$ of the FACH bandwidth. Table 7-9 tabulates the values of a, α and β using this new method for different values of the update interval.

The results in Table 7-10 indicate that with the old way of choosing the parameters, the number of switches per page is high when the backhaul delay is large. The new way of choosing the parameters, enable the adaptive algorithm to enjoy the benefits of LRU algorithm without its penalty.

7.2   Techniques for Improving OPEX/CAPEX of UMTS RAN

In Chapter 6, we discussed how one can determine the bandwidth requirement for the Iub interface between a NodeB and RNC based on the

Quality of Service (QoS) requirement for the voice and data users. The results indicate that the usable utilization for the Iub link is relatively low for data users. Here, we propose a technique[13] that can help to increase this usable utilization at the Iub link via a rate control scheme. In this approach, we control the rate of data offered to the Iub link by changing dynamically how the user data is mapped to the Transport Block Set Size (TBSS) in a TTI. Without rate control, as long as the data present in a particular RLC buffer exceeds the size of the largest Transport Block Set Size, the largest TFB is used to pass the data to the next layer. With the proposed rate control algorithm, the TFB for a particular user is determined based on the current status of all the RLC buffers that have data to transmit. If there are more data than the Iub link can carry at that point in time, then smaller TFBs are used for some of the users such that the aggregate traffic offered remains below the sustainable rate.

Let the threshold number of users that can be supported at full rate be T. If the "equivalent" number of users at peak rate, N_P, exceeds the threshold, then the proposed peak rate control algorithm takes action in order to adjust the rate for each user. Note the use of the term *"equivalent" number of users at peak rate*. For example, suppose that at some instance in time, there are a total of 4 users with non-empty RLC buffers. If 2 of these users are currently transmitting at the peak rate of 64 Kbps and 2 users are transmitting at 32 Kbps, then the "equivalent" number of users at peak rate is roughly 3^{14}.

The Fair Rate Control (FRC) algorithm can be summarized in the flow chart given in Figure 7-9. First, we provide the terms and definitions used:
- T threshold number of users
- NP the equivalent number of users at peak rate
- N the number of users with non-empty RLC buffers
- R peak rate, also referred to as the Radio Bearer Rate, or the full rate
- $RTFj$ the rate that corresponds to TFj (Table 7-8), also referred to as reduced rate
- RS the rate that can be sustained per user at some given point in time
- NL the number of users that are forced to lower rate by rate control

---

[13] This work was done by Dr. S. Abraham, Dr. M. Chuah, Dr. C. Saraydar while they were at Bell Laboratories, and some performance results were published in two conference papers [Sar03a],[Sar03b].

[14] The presence of a TBSS independent FP overhead skews the ratios in reality. So a user at "32 Kbps" is equivalent to slightly more than half a 64-Kbps user.

- *Rate control instance* The time at which either at least one empty RLC

  buffers receives data or at least one RLC buffer is left with no data.
- *Max TFB size*   The largest block the user is allowed to use within the TFS.

At each rate control instance, the number of users that have data to send (N) and the equivalent number of users at peak rate (N_P) are obtained. Based on N, the sustainable rate is calculated as $R_S = R*T/N$ where T is the threshold number of peak rate users. R_S represents the sustainable rate, which should be used by a peak rate user in order to satisfy link capacity constraint. Once the sustainable rate is calculated, the algorithm checks to see whether the equivalent number of peak rate users exceeds the threshold, $N_P > T$. If so, then all users are returned to the peak rate assignment to allow a fair assignment in the subsequent steps of the algorithm. The next step is to identify the maximum rate within the transport format set (R_{TFj}) that is smaller than the sustainable rate (R_S) since only a finite number of rates is specified by the 3GPP standard. Once R_{TFj} has been determined, we calculate how many users need to be restricted to R_{TFj} by the equation in block (1) in Figure 7-9. The calculation in block (1) is an average of rates between N_L users transmitting at R_{TFj} (the reduced rate) and the ($N - N_L$) users transmitting at R (the peak rate). Notice that with this algorithm, at any point in time there are exactly two sets of users: 1) users at full rate, and 2) users at the reduced rate. As stated before, the number of users to be forced to a reduced rate is calculated such that the total average rate remains within the sustainable rate. The next step is to randomly select N_L users and change their maximum peak rate from R to R_{TFj} temporarily. We let $N_L(old) = N_L$. This assignment will remain until potentially being changed in the next rate control instance. In the next rate control instance, N and N_P at that instance are re-established and the sustainable rate recomputed based on the new configuration. If the threshold is exceeded, the same set of procedures is repeated as described earlier in this paragraph.

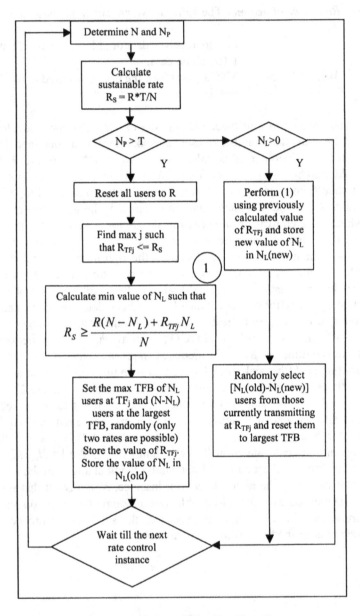

Figure 7-9 Rate Control Algorithm Flow Chart

If the equivalent number of peak rate users is below the threshold, T, then we go through the right-hand branch of the flow chart in Figure 7-9 with the intention of restoring some reduced-rate users back to peak rate as permitted by the newly available bandwidth. If there are currently no reduced-rate users and the equivalent number of users is below the threshold, then we simply wait for the next rate control instance. However, if there is at least one reduced-rate user, we re-compute the users that need to be pushed to a reduced rate by using the expression in block (1). The new computation is based on the previously computed value of the reduced rate (R_{TFj}) and the current conditions (the number of users with data to transmit, the equivalent number of peak rate users, the new value of the sustainable rate). The new number of reduced-rate users is denoted by $N_L(new)$. We then randomly select $N_L(old) - N_L(new)$ users and restore their maximum rate to peak rate.

It should be emphasized that with this algorithm, potentially there are two sets of users at any given point in time, the set at peak rate and the set at reduced rate. Variants of this algorithm, where there are more than 2 rate settings are available, can be considered. However, in such an algorithm, since there are more degrees of freedom, assignment of rates and restoration of rates would be more complex. On the other hand, the advantage would be that the gap between the right-hand-side and the left-hand-side of the expression in block (1) may be smaller.

A variant of FRC, which will be referred to as Unfair Rate Control (URC), determines the number of users that can be supported at full rate and gives only so many users the peak rate. All other users with data to transmit will be given zero rates, i.e., their transmission will be delayed until the load on the Iub link improves. One can think of URC as a special case of FRC where the reduced rate is fixed as $R_{TFj} = 0$. The URC approach performs worse from a user perspective since the user that gets the reduced rate is exposed to longer RLC buffer delays.

Some numerical results of the proposed scheme were presented in [Sar03a], [Sar03b]. They indicate that the proposed scheme improves performance considerably. The FRC performance for web-browsing traffic was studied on a T1 link, for 3 different bearer rates: 64 Kbps, 128 Kbps, and 384 Kbps. The terms and metrics used in the papers were summarized as follow:

Table 7-11 The Transport Format Block Sizes and the Corresponding Bit Rates

Index j	Max TFB size (byte)	R_{TFj} (Kbps)
0	0	0
1	40	16
2	80	32
3	120	48
4	160	64

Table 7-12 List of Parameters Used in the Simulation

TOAWS	30 ms
TOAWE	10 ms
Step Size (Fixed Step Adjust [Figure 7-9])	1 ms
AAL2 Timer_CU	4 ms

- Number simultaneous sessions: The number of users that are in Cell_DCH and URA_PCH modes.

- Frame discard probability (FDP): The probability that a frame arrives at the NodeB at a time too late for transmission to the UE and therefore is discarded.

- Mean and std deviation of Iub delay: Mean and std deviation of the time from the frame is generated at the FP layer to the time the frame is received at the Node-B (includes AAL2/ATM queuing delay and Iub link transmission delay).

- Mean and std deviation of UTRAN delay: Mean and std deviation of the time from frame is generated at the FP layer to the time the frame is scheduled for transmission to the UE.

- Utilization: Ratio of the aggregate bit rate (including all overheads) over the Iub link to the bandwidth of the link.

- Average user bit rate during ON time: The rate at which the user data is transmitted whenever a user has data to transmit.

- Average number users at peak rate: Average number of users that are equivalently transmitting at peak rate (with overheads) at any given point in time.

- Average number of Cell_DCH Users: Average number of users that are either actively sending/receiving data or that are in the viewing mode but the inactivity timer is yet to expire.

- Maximum number of Cell_DCH Users: The maximum number of Cell_DCH users observed at any point in time.

A list of parameters and the values used in obtaining the reported results are given in Table 7-12.

Our studies show that FRC achieves considerable improvement in utilization. Table 7-13 summarizes the results obtained for 64-, 128-, and 384-Kbps bearer services. We present the number of simultaneous sessions that can be supported by the Iub link with a FDP target of less than 0.5%. With no rate control, as long as it has data to transmit, each user is given the highest rate possible through the assignment of the largest block size in the TFS. However, with rate control, at times of congestion on the Iub link, the maximum block size a user can use is restricted, thus resulting in lower rates than the maximum bearer rate. The results in Table 7-13 show that the average user bit rate (during the ON time) with FRC algorithm deteriorates by about 10% as compared to the rate a user gets with no rate control. However, the increase in allowable link utilization in return for the decrease in user bit rate is very significant, e.g., 44% increase from 0.57 utilization to 0.82 utilization for the 64-Kbps bearer service. It can be seen from the table that other bearer services experience similar, and even better increases in the allowable link utilization.

For 64-Kbps service, our airlink analysis indicates that we can support a maximum of 16 cell DCH users per sector. Here, we make no distinction on whether the user is in soft handoff mode or not. From Table 7-13, we observe that a T1 link can support 2 maximally loaded sectors (40/16 > 2) without rate control. With rate control, a T1 link can support 3 maximally loaded sectors (58/16> 3). Similarly, for 128-Kbps service, we have enough power to support a maximum of 8 Cell_DCH users per sector. Without rate control, one T1 link can support 3 maximally loaded sectors (24/8 = 3) with 384-Kbps users, while with rate control one T1 link can support 5 maximally loaded sectors (41/8 > 5). With 384-Kbps service, each maximally loaded sector can support 7 Cell_DCH users. Without rate control, one T1 link can support 1 maximally-loaded sectors (10/7 > 1). With rate control, one T1 link can support 3 maximally loaded sectors (22/7 > 3).

Table 7-13 Numerical Results for a T1 (1536 Kbps) Iub Link

	No Simult. Sessions	Utilization (%)	Avg. User BR during ON time (Kbps)	Avg. No Users at Peak Rate	Avg. No Users in Cell_DCH	Max No. Users in Cell_DCH
64 Kbps						
No FRC	58	57	63.56	10.62	24.85	40
With FRC	85	82	57.15	15.34	37.08	58
Improvement (%)	46.55	44.48	-10.08	44.48	49.23	45
128 Kbps						
No FRC	38	43	126.82	4.03	14.05	24
With FRC	68	73	114.05	6.92	25.17	41
Improvement (%)	78.95	71.56	-10.08	71.56	79.11	70.83
384 Kbps						
No FRC	12	14	347.40	0.46	3.86	10
With FRC	37	42	312.43	1.36	11.67	22
Improvement (%)	208.33	196.24	-10.06	196.24	202.66	120.00

Acknowledgments

The work on Iub interface capacity and rate control algorithms was jointly done with former Bell Lab colleagues Dr. S. Abraham, Dr. C. Saraydar, and Dr. A. Sampath. The model discussed in Section 7.1.4 was proposed by Dr. W. Luo and Dr. C. Chou, from Bell Laboratories.

7.3   Review Exercises

1.   Assume exponential user think time distribution and 64 Kbps service, use the Web traffic model in Table 7-4 and an inactivity timer value of 10 s, evaluate the mean number of times a user will be moved from Cell_DCH to URA_PCH states and vice versa during one Web page download transaction. Then, determine the number of signaling messages generated per Web page download transaction.

2.   Work out the value of E for Solutions 1 and 2 using the example given in Section 7.1.4. Compare the value of E obtained if a gamma distribution with the same mean is assumed for the user think time.

3.   View the UMTS network as a system that receives work in terms of pages, serves the pages, and releases the user. Let λ_A be the new user arrival rate, n be the average number of pages per user, k be the number of available DCH channels, and μ_s be the maximum rate at which a page is completed at a DCH. The necessary and sufficient condition for stability is that the page input rate must not exceed the maximum possible page departure rate which is

$$n \lambda_A < k \mu_\sigma$$

To obtain the statistics of the number of users in the system, the page download delay, we can use the model[15] shown in Figure 7-10.

---

[15] The queueing model described here was proposed by Dr. S. Abraham.

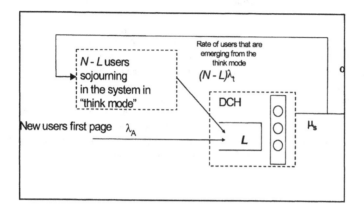

Figure 7-10 Analytical Model for DCH Allocation Scheme

where N is the mean number of users in the system, L is the mean number of users in the DCH mode and λ_t is the reciprocal of the mean think time. We make some simplifying assumptions that both the service time per page and the service time are exponentially distributed. Thus, the users in think mode can be described by an M/M/∞ queue and the DCH queue is a M/M/k queue. For a stable system, show that

$$(N - L)\, \lambda_t + \lambda_A \;= n\, \lambda_A$$

Let d denote the time elapsed from the time the user moves into the DCH queue to the time that the user finishes downloading the page and returns to the think time phase. Then, the soujourn time of a user in the system is given by

$$T = nd + (n\text{-}1)/\lambda_t$$

Using Little's law, show that the number of sojourning users in the system as

$$N = T/\lambda_A.$$

7.4 References

[25.101] 3GPP TS25.101, "User Equipment (UE) radio transmission and reception (FDD)," V3.11.0, July 2002.

[25.402] 3GPP TSG RAN, "Synchronization in UTRAN, Stage 2," TS 25.402 V.3.2.0, June 2000.

[25.427] 3GPP TSG RAN, "Iur/Iub User Plane Protocol for DCH Data Streams," TS 25.427, V.3.3.0, June 2000.

[Abr02] S. Abraham, A. Sampath, C. Saraydar, M. Chuah, "Effect of Timing Adjust Algorithms on Iub Link Capacity," Proceedings of VTC 2002, pp311-315.

[Chou03] C. Chou, W. Luo, "Comparison of battery consumptions with and without URA_PCH state," Bell Laboratories Internal Memorandum, 2003.

[Chu02] M. Chuah, W. Luo, X. Zhang, "Impacts of Inactivity Timer Values on UMTS System Capacity," Proceedings of WCNC, 2002.

[Sar03a] C. Saraydar, S. Abraham, M. Chuah, A. Samptah, "Impact of Rate Control on the capacity of an Iub link: Single Service Case," Proceedings of ICC 2003.

[Sar03b] C. Saraydar, S. Abraham, M. Chuah, "Impact of Rate Control on the capacity of an Iub Link: Mixed Service Case," Proceedings of WCNC 2003.

Chapter 8

CORE NETWORK DESIGN AND TRAFFIC ENGINEERING

8. INTRODUCTION

Figure 8-1 illustrates a logical architecture for a UMTS Network which consists of Terminal Equipment (TE)/Mobile Terminal (MT), UMTS Terrestrial Radio Access Network (UTRAN) and the components that comprise the UMTS Core Network. In a UMTS Core Network, the Serving GPRS Support Node (SGSN) and the Gateway GPRS Support Node (GGSN) are special routers that are used to deliver packet switched services while the Mobile Switching Center (MSC) is used to deliver circuit switched services. In addition, one can find the Charging Gateway Function (CGF) which collects charging information provided by a set of SGSNs and GGSNs and downloads such data to a billing system so that end users can be charged for the services. Home Location Register (HLR), which stores subscribers' profiles, e.g., Access Point Name (APN), the types of services a subscriber buys, is also provided. Often, a Domain Name Service (DNS) server will be available to resolve APNs and neighboring SGSN names. A Dynamic Host Control Protocol (DHCP) server may also be provided to dynamically allocate IP addresses to mobile users.

In this chapter, we first describe in Section 8.1 the procedures that a data user will invoke to activate a packet data session. Then, in Sections 8.2 and 8.3, we discuss the functionalities provided by the two core network elements that provide packet switched services, namely the SGSN and GGSN. Next, we describe in Section 8.4 the details of the GPRS Tunneling Protocol (GTP) tunnel which is used to transport data packets between the Radio Network Controller (RNC) and SGSN, and between SGSN and GGSN. In Section 8.5, we describe a methodology for designing SGSN based on some performance requirements. Typically, SGSN and GGSN is designed to meet certain performance requirements based on some

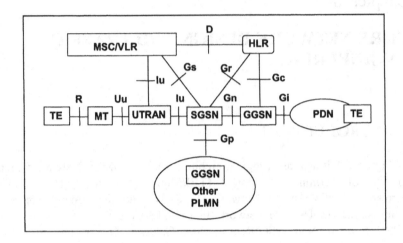

Figure 8-1 Logical Architecture for UMTS Network

assumptions, e.g., the average load generated per active subscriber. Thus, there will be situations where users become active simultaneously and push the system into an overload situation. Therefore, overload control strategy needs to be implemented in the core network elements. We discuss some overload control strategies in Section 8.6. Overload control actions may include rejection of new connection requests as well as dropping of traffic packets. The buffer management scheme implemented inside the core network element decides which packets will be discarded during overload conditions. In addition, both the buffer management and the scheduling schemes in SGSN/GGSN are used to provide differentiated services to the traffic from different QoS classes. We discuss both buffer management and scheduling schemes for SGSN/GGSN in Section 8.7. Last but not least, in the design of the UMTS Core Network, the designers need to determine how the clusters of SGSNs can be homed to regional GGSNs such that the total equipment and transport cost can be minimized. We discuss this topic in Section 8.8.

8.1   Registering and Activating the Circuit/Packet Switched Service

A mobile station needs to register with a UMTS network before it can use any circuit or packet switched services. There are four steps in the registration sequence: (1) The mobile station powers up and scan all the Radio Frequency carriers and search for the strongest cells (2) Based on the identities of the Public Land Mobile Networks (PLMNs) it received from the

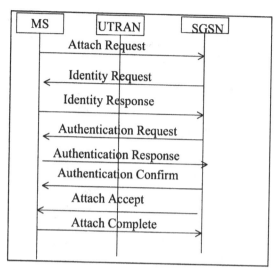

Figure 8-2 GPRS Attach Procedure

cell broadcast channel, it selects a PLMN from the allowed priority list (3) Then, it selects the strongest cell in the selected PLMN that satisfies cell selection criteria and camps onto that cell (i.e., select and monitor page indicator channel and paging channel of the cell, performs cell measurements) (4) After camping on the cell, it performs either an IMSI-attach procedure to 3G MSC/VLR for accessing the circuit-switched services or a GPRS-attach procedure to 3G SGSN for accessing the packet-switched services.

The GPRS attach procedure [23.060] is shown in Figure 8-2. A mobile station (MS) sends an attach request to a Serving GPRS Gateway Node (SGSN). The attach request message contains information such as the MS's identity (e.g., the 15-digit International Mobile Subscriber's Identity [IMSI] or the 32-bit Packet Temporary Mobile Subscriber Identity [P-TMSI]) and the attach type to be executed, and has a typical data size of 50 bytes (including the overhead of SCCP/MTP3 headers). The SGSN will send an identity request message to the MS if the MS uses P-TMSI in its attached request message and has changed a SGSN since the last detached, and the MS will respond with an identity response message. The identity request message is about 20 bytes and the identity response is about 57 bytes (including IMSI of 15 bytes, an authentication triplet of 29 bytes and the overhead of SCCP/MTP3 headers). After receiving the identity response, the SGSN will send an authentication request message to the MS and the MS will respond with an authentication response. Both the authentication request

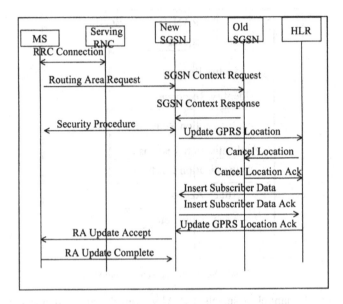

Figure 8-3 Routing Area Update

and authentication response messages are about 40 bytes. The SGSN then sends an authentication confirm message (about 40 bytes) back before sending an attach accept message (about 16 bytes). The attach accept message contains the newly allocated P-TMSI, P-TMSI signature and the radio priority for Short Message Service. The MS then responds with an attach complete message. The IMSI-attach (for circuit-switched service) procedure is similar to the GPRS-attach procedure except that the messages are exchanged between a MS and a MSC/VLR. In addition, the temporary identity assigned by MSC/VLR will be the TMSI.

8.1.1 Routing Area Update

After the GPRS-attach procedure, a mobile station needs to perform a routing area update procedure (shown in Figure 8-3) if this is the new SGSN it is connecting to. A Radio Resource Control (RRC) Connection is established between the MS and the Serving RNC. Then, the MS sends a Routing Area Update Request message to the new SGSN. If an old SGSN exists, then the MS needs to provide the old Routing Area Identity and old P-TMSI information to the new SGSN. The new SGSN sends a context request to the old SGSN so that it can get the MS's IMSI. After hearing a response from the old SGSN, the new SGSN may invoke security procedure

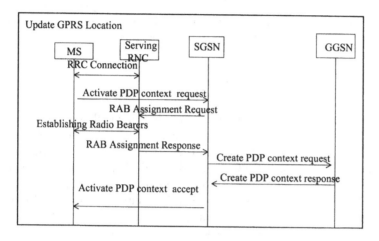

Figure 8-4 PDP Context Activation

to authenticate the MS. The new SGSN informs the HLR of a change in SGSNs. The HLR sends a "Cancel Location" order to the old SGSN to detach the MS's IMSI. The old SGSN acknowledges cancellation with HLR. The HLR sends subscription data to the new SGSN. The new SGSN acknowledges receipt of this subscriber data before HLR acknowledges the GPRS location update to the new SGSN. The new SGSN then sends a Routing Area Update Accept message which includes the new P-TMSI to the MS. The MS then responds with an RA Update Complete message acknowledging receipt of the new P-TMSI. The new SGSN can release the signaling connection to the serving RNC and the serving RNC may release the RRC connection if it is not needed to maintain a CS domain connection.

8.1.2 Activating a Packet Data Session

The mobile station needs to perform a PDP context activation procedure to initiate a data session. The activate PDP context procedure is as shown in Figure 8-4. The MS establishes an RRC connection with the Serving RNC. Then, the MS sends an activate PDP context request message to the SGSN. The SGSN requests the setting up of Radio Access Bearers (RABs) with the radio access network via a RAB assignment request message to the Serving RNC. The Serving RNC, the Node-B and the MS set up the RABs. Then, the Serving RNC sends a RAB assignment response message back to the SGSN and a GTP-U tunnel between the Serving RNC and the SGSN is set up. The SGSN sends a create PDP context request to the GGSN to set up a PDP context since a connection to an external Packet Data Network is required.

The GGSN allocates a PDP address (dynamic IP assignment) and sends a create PDP context response back to the SGSN. The SGSN then sends an activate PDP context accept message to the MS with the assigned PDP address.

8.1.3   Receiving a CS Domain Call

Here, we describe a scenario that shows how a mobile station can receive a circuit-switched call assuming that an existing packet switched connection is already established. The whole process is captured in Figures 8-5 and 8-6. First, the PSTN contacts the Gateway MSC (GMSC) in the home network of the mobile station with an ISUP initial address message (see Figure 8-5). The GMSC then asks the HLR to send the routing information (SRI). The HLR asks the visited VLR to provide the roaming number (PRN) so that the GMSC can forward the call. The visited VLR replies with the routing number to the HLR in the PRN acknowledgement. The HLR then forwards the routing number to the GMSC in the SRI acknowledgement.

The GMSC then forwards the call to the routing number which is received by the visited MSC/VLR (see Figure 8-6). The "page MS" message is sent to the SGSN over the Gs interface. The SGSN sends a page request to the SRNC over the RANAP signaling connection with the Core Network Domain Indicator (CNDI) set to "Circuit Switched" (CS). The SRNC relays the paging request to the MS over an existing RRC connection. The MS responds with a page response message over the RRC connection. The SRNC then relays the page response to the MSC/VLR over the RANAP connection. The CallSetup message is sent to the MS from the MSC/VLR with the bearer capability information. The MS responds with the Call confirm message to the MSC/VLR. The MSC/VLR then asks the SRNC to allocate the traffic channel. The MS is notified of the allocated traffic channel.

8.2   SGSN

A SGSN is analogous to an MSC in the sense that an MSC performs call management for voice services while a SGSN performs session management for packet data services. A SGSN communicates with an HLR to register a mobile and obtain a profile of authorized services. A SGSN also performs authentication and service authorization, creates and maintains connections to packet services, etc.

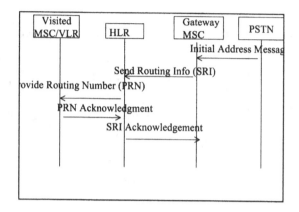

Figure 8-5 Retrieving Routing Information

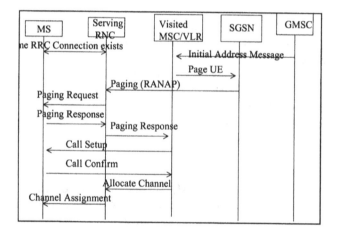

Figure 8-6 Simplified MS Terminated Circuit Switched Call Scenario

The functions supported by SGSN are:
- Registration of a mobile with its HLR
- Authentication of a mobile
- Authorization and admission control of packet services for a subscriber
- Packet session establishment, maintainence, and teardown
- Supports PDP context activation
  - PPP based PDP contexts
  - PPP relay services

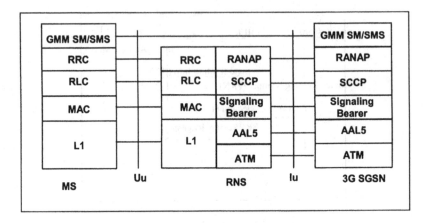

Figure 8-7 Control Plane Between RNS and 3G SGSN [23.060]

    o   IP-based PDP contexts
- Provides packet switching and routing for UMTS/GPRS packet services
- Contains database of subscribers currently being served
- Supports GPRS attach procedures
- Supports GPRS Tunneling Protocol (GTP) used across the Iu-PS interface and Gn/GP interface
- Collection of charging data

The control signaling handled by SGSN supports the general functions of Packet Mobility Management (PMM), Session Management (SM), and Short Message Service (SMS). PMM includes mobile attach (analogous to MSC registration), HLR interactions for authentication and service authorization, and routing area tracking. SM supports the creating and maintaining of packet data sessions, each of which is called a Packet Data Protocol (PDP) context. So, the SGSN must create and maintain a record for each active PDP context record. The PDP context is analogous to a Mobile Call Register (MCR), which is a record of the state of a circuit call.

The SGSN is typically connected to several RNCs via an ATM network. Between the SGSN and its associated RNCs, separate ATM PVCs are used to carry RAN-side control messages and end user traffic. RAN-side control messages include messages between RNC and SGSN and messages between the mobile and SGSN. The protocol stack for the control plane between a radio network system (RNS) and 3G SGSN is shown in Figure 8-7. All

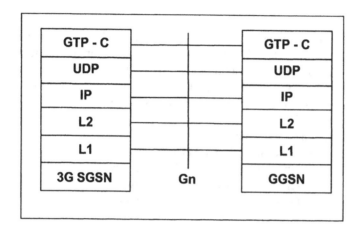

Figure 8-8 Control Plane Between 3G SGSN and GGSN [23.060]

RAN-side control messages are carried within a protocol layer called the RAN Application Part (RANAP). RANAP is carried over SCCP over Broadband SS7 (SCCP/MTP-3B/SAAL/ATM – see the 25.41x series of 3GPP Technical Specifications). For the most part, RANAP uses SCCP connection-oriented service. However, there are a few cases where RANAP uses SCCP connectionless service, e.g., paging.

The SGSN has narrowband SS7 interfaces to support signaling with HLRs, MSCs, Short Message Service-Service Centers (SMS-SCs) and SCPs. For all these end points, the lower layers of the stack are SCCP(connectionless)/MTP3/MTP2. For signaling with HLRs and SMS-SCs, the application layer is Mobile Application Part (MAP) over TCAP. For communication with MSCs the application layer is BSSAP+. The application layer for CAMEL is called the CAMEL Application Part, or CAP.

The SGSN is connected to GGSN via an IP-based network. The control messages between SGSN/GGSN are carried over a GPRS Tunneling Protocol (GTP) tunnel. The protocol used is the GTP-Control Protocol (GTP-C). The protocol stack for the control plane between the 3G SGSN and GGSN is shown in Figure 8-8.

8.3   GGSN

GGSN is the gateway node between a GPRS/UMTS backbone system and external networks. GGSN is the anchor point for the duration of the whole data session. If the mobile moves to a different routing area and thus to a different SGSN, SGSNs and GGSN coordinate the move such that the same GGSN is still providing the connection between the mobile user and the outside world.

The functions provided by GGSN are:
- Supports relevant GPRS/UMTS interfaces:
  - o   Gn – with a SGSN in the same network
  - o   Gp – with a SGSN in other networks
  - o   Gi – IP interfaces to outside world
  - o   Ga – GTP' interface with CGF for billing
- Supports GTP
- Supports legal interception
- Interacts with HLR to locate and determine routing to destination mobile
- Provides Static/Dynamic IP address assignment
- Routing of data packets
- Authentication Authorization, Accounting
- Traffic control and management, e.g., filtering out irrelevant packets and performing congestion control.
- Service performance measurements and accounting data collection.

GGSN terminates the GPRS Tunneling Protocol (GTP) tunnel [29.060] running over an IP based backbone between a SGSN that the user is currently attached to and GGSN. If the GGSN and the SGSN supporting a GTP tunnel belong to the same operator, the interface between them is referred to as the Gn interface. If they belong to different operators, then that interface is referred to as the Gp interface. The Gp interface requires more advanced security mechanisms. GGSN is connected to the Charging Gateway Function via the GTP' (a protocol used to carry charging data).

8.4   GPRS/UMTS GTP Tunnel

The protocol used for the tunnel between the SGSN and GGSN is the GPRS Tunneling Protocol (GTP). GTP is analogous to IETF tunneling protocols such as Layer Two Tunneling Protocol [RFC2661], but with additional capabilities to support mobility. GTP is separated into a control plane (GTP-C) and a user plane (GTP-U). GTP-C is used to set up, maintain, and tear down user data connections, while the user data is sent over GTP-U.

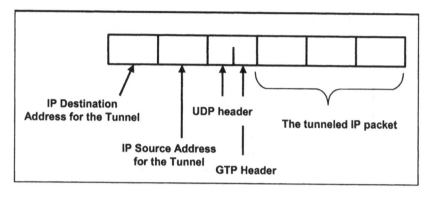

Figure 8-9 GPRS Tunneling Protocol Packet Format

All GTP-C and GTP-U messages are transported over UDP/IP. In addition, the standard mandates GTP-U on the RAN side to be carried over UDP/IP/AAL5/ATM. GTP is described in a 3GPP document, TS 29.060. User packets are encapsulated within GTP-U "tunnels" between the RNC and SGSN and between the SGSN and GGSN. So from the point of view of the user packets, GTP-U provides a layer 2 tunnel between the RNC and SGSN. For example, GTP tunnels can be handed off from one RNC to another and from one SGSN to another without interrupting the user data session.

The GTP packet format is shown in Figure 8-9. The original IP packet is encapsulated within a UDP/IP packet. In addition, a GTP header is inserted. The outline of the GTP header is shown in Figure 8-10. The first byte of a GTP header contains the Version Number and some option flags. The second byte of the GTP header indicates the Message Type. This is followed by a 2-byte length field. Then comes the 4-byte tunnel identifier. After the tunnel identifier, there is a 2-byte sequence number. After the sequence number is the N-PDU number. This is followed by a next-extension header type.

Octets	Bits							
	8	7	6	5	4	3	2	1
1	Version			PT	(")	E	S	P N
2	Message Type							
3	Length (1st Octet)							
4	Length (2nd Octet)							
5	Tunnel End Point Identifier (1st Octet)							
6	Tunnel End Point Identifier (2nd Octet)							
7	Tunnel End Point Identifier (3rd Octet)							
8	Tunnel Endpoint Identifier (4th Octet)							
9	Sequence Number (1st Octet) (1)							
10	Sequence Number (1st Octet) (1)							
11	N – PDU Number (2)							
12	Next Extension Header Type (3)							
(") – This bit is a spare bit.								
(1) – This field is evaluated only when S bit is set.								
(2) – This field is evaluated only when PN bit is set.								
(3) – This field is evaluated only when E bit is set.								

Figure 8-10 Outline of GTP Header [29.060]

8.5   Capacity Sizing of SGSN/GGSN

Typically, a SGSN/GGSN is designed to meet some performance requirements, e.g.:

- A SGSN can maintain N active PDP contexts without IPSEC and M active PDP contexts with IPSEC. A traffic model needs to be assumed so that one can decide how much throughput a SGSN needs to sustain with these many active PDP contexts. For example, we can assume that each PDP context generates 6.2 Kbps traffic and each user may generate 4 PDP contexts. We can further assume that 50% of the active PDP contexts generate traffic simultaneously. Such assumptions allow the vendor to determine the sustained throughput that need to be supported by each SGSN.

- A SGSN can sustain a throughput of S Mbps assuming an average packet data size of P-bytes. For example, a medium size SGSN may be designed to support 600 Mbps assuming an average packet size of 500 bytes and that each PDP context is assumed to generate about 6.2 Kbps traffic.

- Ideally, a SGSN should be able to be connected to as many RNCs (for SGSN) or SGSNs (for GGSN) as possible. However, in real life, due to memory constraints, a vendor may limit the maximum number of RNCs that can be connected to a SGSN and the maximum number of SGNS that can be connected to a GGSN.

In this subsection, we describe a methodology for estimating the signaling/bearer load for a SGSN. Similar methodology can be used for the capacity sizing of GGSN.

8.5.1 Signaling Load Estimate

Let the total signaling load processed by a SGSN during busy hour to be S_{BH}. S_{BH} is the sum of signaling due to mobile related activities, $S_{M,BH}$, and signaling due to non-mobile related activities, $S_{NM,BH}$, e.g., keep-alive messages during a busy hour.

$$S_{BH}(Kbps) = S_{M,BH} + S_{NM,BH} \qquad (8\text{-}1)$$

The signaling loads due to mobile-related activities include the signaling load generated by all mobiles within the coverage area of a SGSN for the attach/detach procedures; intra-SGSN, inter-SGSN, and periodic routing area updates; PDP context activations/deactivations, Radio Access Bearer (RAB) setup/teardown procedures, etc. The signaling traffic due to non-mobile related activities is mostly vendor specific and hence will not be discussed further here.

The signaling load generated by mobile related activities is a function of the mobile speed and some design parameters such as PDP context lifetime, RAB inactivity timer, etc.

$$S_{M,BH}(Kbps) = S_{attach/detach} + S_{RAU} + S_{SMS} + S_{Session} \qquad (8\text{-}2)$$

$$S_{RAU}(Kbps) = S_{IntraSGSN\,RAU} + S_{InterSGSN\,RAU} + S_{Periodic\,RAU} \qquad (8\text{-}3)$$

$$S_{session}(Kbps) = S_{PDP} + S_{paging} + S_{RAB} \qquad (8\text{-}4)$$

where $S_{attach/detach}$ is the total signaling traffic generated by all mobiles supported by the SGSN during the attach/detach procedures, $S_{IntraSGSN\,RAU}$,

$S_{InterSGSN\ RAU}$, $S_{Periodic\ RAU}$ are the total signaling generated by all mobiles when they perform intraSGSN, interSGSN and periodic routing area updates during busy hour. S_{SMS} is the signaling traffic generated by SMS-related activities; $S_{session}$ is the total signaling traffic generated by session related activities which consists of three major components, S_{PDP}, S_{paging} and S_{RAB}. S_{PDP} is the total signaling traffic generated by all mobiles for performing the PDP context activation/deactivation procedures; S_{paging} is the total signaling traffic generated by SGSN in paging the relevant mobiles, e.g., PDU notifications; S_{RAB} is the total signaling traffic generated by all mobiles for performing the RAB setup/teardown procedures.

Similarly, by changing all the S to M where M denotes the number of messages per second that are generated by these various activities, we have

$$M_{M,BH}\ (messages/s) = M_{attach/detach} + M_{RAU} + M_{SMS} + M_{Session} \qquad (8\text{-}5)$$

To estimate the signaling load generated by the attach/detach procedures, we need to count the number of messages invoked per procedure and make an assumption on how frequently each subscriber will perform the attach/detach procedures. For an example, in the attach procedure (refer to Figure 8-2), we have four downlink messages (from SGSN to MS) and four uplink messages (from MS to SGSN). Including the overhead of SCCP and MTP3, the attach request, and the identity response message exceeds 47 bytes and hence it takes two ATM cells to transport such messages. The rest of the messages can be transported using one ATM cell each. This explains the number put in the first row of Table 8-1 under attach procedure. Similar steps can be taken to fill up the rest of the rows in Table 8-1.

To estimate the signaling load generated by the routing area updates, we need to make certain assumptions:

- An intra-SGSN routing area update (RAU) happens when there is a change of routing area but no change in SGSN. An inter-SGSN RAU happens when there is a change of routing area and SGSN. In our estimate, we assume that each RNC spans a different routing area. Thus, a change of RNC will trigger either a intra-SGSN RAU or an inter-SGSN RAU. This is a worst case assumption.

- The following formula is used to estimate the RNC/SGSN crossover rate.

The rate of boundary crossover rate, N, by mobiles moving with a velocity of V in a cellular system with a cell perimeter of L assuming uniformly distributed motion is given by

$$N = (VL \rho)/\Pi \tag{8-6}$$

where ρ is the mobile density.

For RNC/SNSN crossover rate, L will be the perimeter of the RNC/SGSN coverage area.

- Periodic RAUs are performed only by attached users that are in PMM-idle mode (i.e. they do not have an active session). Thus the rate of periodic RAUs can be estimated using the following formula:

$$\text{Periodic RAUs} = H_2/P_{refresh} \tag{8-7}$$

where H_2 is the total number of PMM-idle subscribers within the coverage area of a SGSN. H_2 can be estimated by estimating the number of PMM-connect users based on the services they use, e.g., web-browsing, streaming video, email, etc. By using the M/G/∞ and the relevant session arrival/exit model for each service, one can determine the number of PMM-connect subscribers for each service. H_2 is the difference between the total number of attached subscribers and the total number of PMM-connect subscribers.

- $S_{session} = \Sigma S_{session,i}$, where $S_{session,i}$ is the total signaling load generated by session-related activities for each service. Such load is a function of the traffic model used for a particular service, e.g., web-browsing and some design parameters, e.g., PDP context lifetime, RAB idle timer, etc. For example, during a Web-browsing session, if the RAB idle timer is set too small, RAB will be torn down during page reading time and RAB setup procedures need to be invoked every time a new page arrives.

By estimating the message size of each procedure and counting the number of messages that need to be sent across Iu-Ps or Gn interfaces, we come up with Table 8-1.

We provide some numeric numbers based on a simple scenario where all the data users use only Web-browsing application.

Assume that each SGSN is designed to support 500,000 subscribers and that 70% of these subscribers use a Web-browsing service. We also assume the Web-user subscriber's activity factor is 0.17. With these assumptions, there will be 58275 attached Web users (500,000*0.7*0.17). We assume that each subscriber performs one attach/detach per hour. Then, using the information from the first two rows of Table 8.1, we can determine the total ATM traffic generated across the Iu interface by the Web-users performing attach/detach procedures is:

Downlink: 34.3 Kbps
        which is calculated as ((4 + 1)*53*8*58275/3600/1000)Kbps.
Uplink: 48.0 Kbps
        which is calculated as ((6 + 1)*53*8*58275/3600/1000)Kbps.

We further assume that there are 625 NodeBs per SGSN, and that the coverage area of a SGSN is five times that of the coverage area of an RNC. We also assume that the perimeter of each NodeB is 8 km. Assuming that the coverage area is a circle, one can compute that the perimeter of the RNC is about 89.4 km and that for the SGSN is about 200 km. Assuming that the velocity is 60 km/hr (e.g., all Web-browsing users are riding inside the trains), then the SGSN cross over rate can be computed using Equation (8-6) to be about 24.7/s since $\rho = 23.3 = 58725*16/(625*L^2)$ where L is the perimeter of a NodeB and taking the simplification that the coverage area is $(L/4)^2$. The intra-SGSN routing update rate can be determined to be 30.6/s (this is the difference between RNC crossover rate and SGSN crossover rate assuming that each routing area is as small as the coverage of an RNC). To compute the RNC crossover rate, again Equation (8-6) is used and the perimeter of RNC is used rather than the perimeter of the SGSN.

Table 8-1. Message Size/Number of Messages per Procedure Across Iu-PS/Gn Interfaces

	Iu Interface				Gn Interface			
	No of ATM cells		No of msgs		No of ATM Cells		No of msgs	
	Dnlink	Uplink	Dnlink	Uplink	Dnlink	Uplink	Dnlink	Uplink
Attach	4	6	4	4				
Detach	1	1	1	1				
Activate PDP context	7	7	3	3	2	1	1	1
Deactivate PDP context	1	1	1	1	1	1	1	1
RAB setup	2	2	2	2				
Service request	3	4	2	3				

The following Web-browsing traffic model is used to compute the signaling traffic generated by the various procedures:

- Average number of Web pages/session = 20
- Average download/upload size = 64400/660 bytes
- Average download/upload IP packet size = 800/100 bytes
- Average page reading time = 40 s.
- Average number of Web pages per session = 20
- Average downlink/uplink throughput = 30 Kbps/16 Kbps

Note that the above traffic model is obtained by analyzing traffic traces of a wireless system with similar airlink peak/average bandwidth characteristics and is used only as an example here. From the above traffic model, one can estimate the session duration to be about 21.7 minutes. The data activity factor, which is the ratio of the RAB active time over the session holding time is computed to be 31%.

Assuming that the average page reading time is 40 s and a RAB inactivity timer of 120 s means that there is no need to invoke many RAB assignment procedures during a Web-session. We first compute how long each Web-session takes using the Web-browsing traffic model. We assume that it takes 1 s for channel access, 5 s for RAB assignment procedure. Coupled these numbers with a 17 s for transporting the downlink traffic, 3.3 s for transporting the uplink traffic, and a 40 s page reading time, then the average Web-session time will be $(1+5+20.3*20+19*40+120) = 1292$ s. The fastest rate at which each subscriber will perform a PDP context activation procedure is to assume that the user performs a deactivate PDP context procedure after a web-session ends and then immediately starts a new data session again. So, the estimated downlink/uplink signaling traffic for PDP activation procedures across the Iu interface can be computed to be $(10*53*8*58275*1292) = 190$ Kbps (downlink) / 190 Kbps (uplink) respectively using the numbers provided in the third and fourth rows of Table 8-1.

There are several design parameters that need to be set appropriately to avoid generating excessive signaling traffic, e.g., the RAB idle time. If we set it to too low a value, the page reading time will exceed the RAB idle time frequently and hence RAB will be torn down during the page reading time and any new page request will require a new RAB setup procedure. Another factor that affects the signaling load is the user's mobility rate. If the user is moving very fast, its mobile terminal tends to cross the routing area/location area more frequently and hence can generate more signaling load. Different services may generate different signaling load when the PDP context

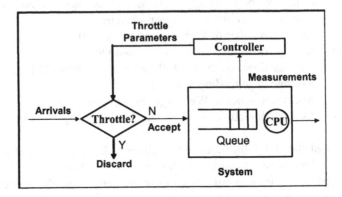

Figure 8-11 Generic Overload Control System

lifetime is set differently. When the PDP context lifetime expires (i.e., no new data session is activated by the same user after this context lifetime timer expires), the PDP context will be deleted from both SGSN/GGSN. If a new data session arrives after the PDP context is deleted, additional signaling messages are required to set up new GTP tunnels etc. Consider the voice-over IP (VoIP) example. If the average call holding time is short, and the PDP context life time is set at 30 minutes, then the UMTS network can keep the user's context information in SGSN/GGSN for a while so a new VoIP call does not need to invoke all the procedures needed for establishing a VoIP call from a cold-start. Only RAB setup procedures need to be performed to allow a new VoIP call to proceed.

8.6   Overload Control Strategy

Packet services generate bursty traffic. Designing a system based on worst case scenario is very costly. Thus, most of the design is based on the estimated average signaling and bearer load. However, this means that occasionally when too many users are active simultaneously, the system will find itself in a congested state. Thus, the system needs to have overload control strategies at various points within its network.

A typical overload control system is shown in Figure 8-11. In a generic overload control system, arrivals will normally be accepted into the system when the measurements of the system indicate that the system load is below the throttle threshold. Whenever the system load becomes congested, overload control actions will be taken. One possible overload control strategy is as follows:

1. Important resources, e.g., CPU usage, buffer usage, and memory usage are periodically sampled. The resource utilization is averaged and the averaged value is compared to a high watermark. If it exceeds the high watermark, a timer is triggered. Anytime, the averaged value drops below the high watermark, the timer is reset.

2. If the average value stays above the high watermark for a certain period of time (until the timer expires), then overload control actions will take place.

3. Once overload control actions take place, the averaged value has to drop below a low watermark for a period of time before the control actions will be stopped.

4. Overload control actions may be rejection of new connection requests, different packet dropping strategies, etc.

Besides requiring a good algorithm to detect overload situation, well-designed load-shedding algorithms are also required for carrying out overload control actions. One specific load-shedding algorithm designed by a Bell Laboratories researcher Dr J. Kaufman is described in [Ka01]. A brief description of the algorithm is given below.

Let T be the measurement and control interval. Let φ denote the fraction of packets/requests allowed to enter into the system during the $(i+1)^{th}$ interval. Let $\varphi_0 = 1$ and φ_i be always constrained within the interval $[\varphi_{min}, 1]$ where φ_{min} is a small but non-zero number which prevents the throttle from discarding all incoming packets/requests. At the end of the ith measurement interval, the load estimate, ρ_i is available and we calculate $\phi_i = \rho_{target} / \rho_i$, where ρ_{target} is the target maximum utilization allowed by the server which is chosen to ensure that the incoming packets/requests are served within a reasonable time. If $\rho_i = 0$, we set $\phi_i = \phi_{max}$ where ϕ_{max} is a large number whose precise value is unimportant. After ϕ_i has been computed, the throttle ratio for the next interval, denoted by φ_i is given by $\varphi_i = \varphi_{i-1} \cdot \phi_i$. Since φ_i must be truncated to lie in the interval $[\varphi_{min}, 1]$, one can rewrite the above as

$$\varphi_i = \max \{ \min \{ \varphi_0 \; \Pi_1^i \; \phi_j \; , 1 \}, \varphi_{min} \}$$

The choice of the measurement and control interval, T, is important for it affects the false alarm rate of the overload control strategy. The interval needs to be short so that control action can respond quickly to any abrupt overload event but it needs to be sufficiently long so that the variance of the utilization estimator is not excessive to reduce the false alarm rate. For the unbiased utilization estimate employed in the above overload control scheme, the central limit theorem suggests that the T s statistical estimates are essentially normally distributed. In addition, the standard deviation of this estimator depends primarily on the average utilization and the length T of the detection and control interval. More specifically, it is proportional to the average utilization and inversely proportional to the square root of the length of the detection and control interval.

Consider the following example[16] : assume that the control and detection interval is chosen to be 10 s and that the standard deviation of the utilization estimator from measurement is approximately 0.05 when the utilization is in the mid 60% range. Assuming stationary traffic and an average utilization of 65%, then the probability of a false alarm (assuming $\rho_{max} = 0.8$) is approximately 0.0013 (a one-sided 3 σ excursion of a normally distributed random variable). This corresponds to one false alarm every 2 hours. To reduce the false alarm rate, one needs to reduce the variance of the utilization estimator. In a stationary traffic environment, halving the standard deviation can shift the 1 false alarm per hour point from a utilization of 66% to 73%. Unfortunately, halving the standard deviation would require quadrupling the detection and control interval from 10 s to 40 s. Because 3–5 intervals of control action is typically require to respond to a significant traffic overload, this would mean that it may take 2–3.5 minutes to respond to an overload situation which in most cases are considered to be too long and not desirable. What one can do is to use a 10 s detection and control interval but takes action only when there are two or three successive excursions beyond ρ_{max}. If the false alarm rate is α when action is taken immediately when the monitored variable exceeds ρ_{max}, then the approach of taking an action only after two successive excursions can reduce the false alarm rate to about α^2. A minor price to pay is that the control action requires an additional of 10–20 s to take place.

There are several ways SGSN/GGSN can perform overload control. For an example, both SGSN/GGSN can perform signaling overload control by dropping GTP-C request messages according to some algorithms when either the network resource consumption exceeds a certain threshold or the CPU consumption of their main controllers exceeds say 85%. Both SGSN/GGSN can also perform traffic overload control by shedding traffic from PDP contexts with lower QOS requirements first. Before shedding traffic from the existing PDP contexts, GGSN shall first deny services to any incoming requests for establishing new PDP contexts.

---

[16] The above example was provided by Dr. J. Kaufman in an internal Bell Laboratories document.

8.7 Scheduling/Buffer Strategies

Figure 8-12 shows a data connection between a mobile user and an internet host over a UMTS network. The GGSN is the gateway between the internet and the UMTS network. It is the entry point through which all packet switched traffic bounds for the UMTS mobiles is routed. Usually, one GGSN may serve many UMTS packet data users that span multiple Radio Network Controllers. Thus, GGSN is the appropriate point in the UMTS network where traffic management is applied such that traffic overload condition will not occur within UTRAN and also to ensure that the users' QoS guarantees that are negotiated at connection setup can be sustained.

The first generation of UMTS equipment often does not support full-blown QoS features. It only offers best effort service and hence only First Come First Serve and Tail Drop buffer management schemes are typically used in SGSN/GGSN. Since the UMTS standard was originally designed to support multimedia sessions, both SGSN/GGSN design needs to provide QoS features ultimately. In this section, we explore how different scheduling and buffer management schemes affect the end-to-end performance of IP multimedia sessions. We explore how scheduling algorithms such as Round-Robin and Weighted Round Robin and buffer management schemes such as Weighted RED [WRED] affect the throughput and packet losses of interactive class (e.g., HTTP), and best effort class traffic (e.g., FTP).

8.7.1 Scheduling Algorithms

Scheduling dictates how a packet departs from each queue, typically at the output interface of a router towards the next router or host, or potentially from one card to other queuing points within the same router. Traditional routers have only a single queue per output link interface. So, the scheduling task is simple – packets are pushed out of the queue as fast as the underlying link allows. In routers that provide differentiated treatments to packets from different users, e.g., Class Based Queue [cbq] architecture, each interface has a scheduler that shares the output link's capacity among the queues associated with the interface. Link sharing is achieved by appropriately scheduling when and how frequently packets are transmitted from each queue.

Because a packet's traffic class dictates which queue it is placed in, the scheduler is the ultimate enforcer of relative priority, latency bounds, or bandwidth allocation between different classes of traffic. A scheduler can establish a minimum bandwidth for a particular class by ensuring that

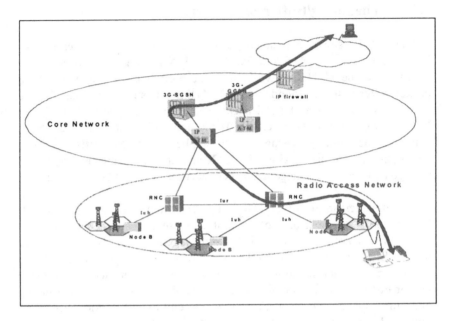

Figure 8-12 Packet Data Connection via UMTS Network

packets are regularly transmitted from the queue associated with that class. A scheduler can also provide rate shaping (imposing a maximum allowed bandwidth for a particular class) by limiting the frequency with which the class's queue is serviced. Depending on a scheduler's design, it might impose both upper and lower bandwidth bounds on each queue or impose upper bounds on some queues and lower bound on others.

Scheduling algorithms are usually a compromise between desirable temporal characteristics and implementation simplicity. Every scheduler design has its own particular service discipline i.e. the manner in which it chooses to service queues. The simplest schedulers focus on servicing queues in some predictable sequence (whether fixed or configurable). More advanced schedulers allow relative or absolute bandwidth goals to be set for each queue, and they continuously adapt their servicing disciplines to ensure the average bandwidth or latency achieved by each queue is within configured bounds.

Some common scheduling algorithms are:
- First In First Out (FIFO): All packets are served according to the times they arrive. Those that arrive first will be served first.

- Strict Priority: High-priority queues will be served first. Low priority traffic may face starvation problem.
- Round-Robin/Weighted Round-Robin scheduling: Queues are serviced in round-robin fashion, the service time at each queue is a function of the weight assigned to each queue. If all the weights are the same, the scheme is called Round-Robin. If the weights are different, the scheme is called Weighted Round-Robin. There are other variations of Weighted Round-Robin schemes. Interested readers can refer to [Cha99] for further discussions.
- Weighted Fair Queueing: A minimum rate is guaranteed per class and the service time of each queue is a function of the packet size and of the queue weight. Current service time is updated everytime a packet is sent.
- Class-Based Queueing [Flo95]: A maximum allocated rate for each class (can be organizational category) is configured. For example, one may configure that all traffic from the engineering department consumes 70% of the access link bandwidth while the traffic from marketing department consumes 30% of the access link bandwidth. Within each class, we can further constrain how the allocated bandwidth can be shared among different applications. For example, we may further constrain that FTP traffic cannot consume more than 10% of the bandwidth allocated to the engineering department.

8.7.2 Buffer Management Schemes

A queue's occupancy increases when the offered load (traffic arrival rate) exceeds the rate at which the scheduler is draining packets from the queue. Because the rate at which any particular queue is drained depends on how its scheduler reacts to traffic from other queues contending for access to the output link, occupancy can be considered to reflect the level of congestion currently being experienced at the queue's output interface.

Reducing a queue's occupancy requires some method of triggering congestion-avoidance behavior in the transport protocols used by the flows passing through the queue. Because a finite delay occurs before any transport protocol can react to congestion within a router, queue management must deal with two basic type of congestion:

- Transient congestion, occurring over time periods shorter than the reaction time of congestion-avoiding transport protocols

- Long-term (average) congestion, resulting from the approximately steady-state rates of all the flows passing through the queue.

Transient congestion is caused by short, correlated bursts of traffic from one or more flows. In general a router does not want to drop packets from a burst unless absolutely necessary. However there is always the chance that a burst will fill the queue, at which point packet drop is the only option available.

A queue manager needs to continuously supply a feedback (in terms of some function of queue occupancy) to transport protocols to keep the long-term occupancy down. In principle you can apply feedback in two ways:

- In-band marking of packets
- Dropping of packets

In-band marking of packets can be used if the transport protocol can react to the receipt of marked packets by initiating congestion avoidance. In practice, packet dropping is the preferred approach in IP networks because TCP uses packet losses to trigger its congestion avoidance behavior. Packet dropping also has a beneficial side effect of immediately reducing the downstream load.

Several schemes that can be used to drop packets during congestion period. Two simple schemes are the drop from front and tail drop schemes. For the drop from the front (DFF) scheme, the packet at the head of the queue is dropped while for the tail drop scheme, the latest arrival is dropped when the queue size exceeds certain threshold. DFF expedites TCPs congestion-avoidance behavior – a "packet loss" event at the head of the queue is noticed sooner than one at the tail of a queue that may already have a serious level of backlogged packets. However, DFF has no impact on non flow-controlled traffic.

Weighted RED [WRED] has been proposed as a queue management scheme to provide differentiations to different flows. WRED calculates a single average queue size based on the total number of packets seen. Different classes of traffic are then dropped using different thresholds. Assume that we have 3 different classes, namely Red, Yellow, and Green. The red packets will be dropped with probability equal to the slope value when the average queue size exceeds Pmin(red). When the total queue size exceeds Pmax(red), all red packets will be dropped. The different Pmin/Pmax values chosen for the different classes creates the three main

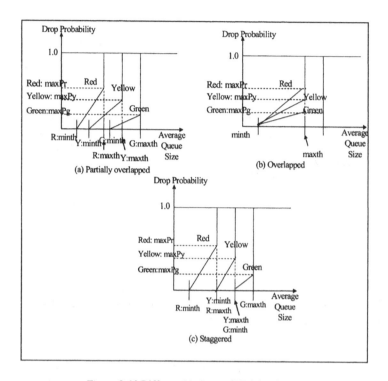

Figure 8-13 Different Variants of Weighted RED

variants of Weighted RED as shown in Figure 8-13. There are other variants of the Weighted RED scheme, e.g., using the mean individual class queue size rather than the total queue size.

8.7.3 Performance Evaluations of Different Scheduling/Buffer Management Schemes

In this section, we discuss how one can compare the different scheduling/buffer management schemes in a GGSN. One can use analytical approach or resort to simulations. Analytical approach sometimes is not tractable or complex. So, here, we present a simulation approach. We implemented the UMTS RLC/MAC as a new module in NS simulator [NS] and use the enhanced simulator to study the impacts of RLC/MAC on the end-to-end performance in this section. From these UMTS RLC/MAC studies, we observe that each mobile will experience an end-to-end TCP throughput of about 45 Kbps when it is assigned a 64 Kbps dedicated airlink channel with an airlink frame error rate of about 10%, and the maximum

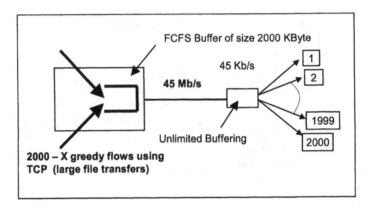

Figure 8-14 Network Configuration for Simulation Studies

number of retransmissions is set to 6. To speed up the simulation study, we replaced the detailed RLC/MAC module with a simple link connecting any mobile to a NodeB with a link throughput of 45 Kbps. Figure 8-14 shows the simulated network we used for the experiments to study the effects of scheduling/buffer management schemes on the end-to-end performance. We assume that the GGSN under study supports 2000 active users. Some of the users are running FTP applications while the rest are running HTTP applications. The link between SGSN and GGSN is assumed to be 45 Mbps and the buffer size at GGSN is assumed to be 2 Mbytes. The throughput of a HTTP session is defined to be the ratio between the total downlink bits received during the session divided by the session time (excluding the viewing time).

We first study the end-to-end throughput of a mobile user when FCFS service discipline is used at the GGSN[17]. Table 8-2 tabulates the results as the number of HTTP and greedy sources are varied. The results indicate that HTTP flows are severely impacted by the FTP greedy flows. Packets from greedy flows whose windows grow to large numbers occupy most of the buffer space. Packets from HTTP flows are dropped. Thus the throughput for HTTP source during busy period is around 50% of the throughput of the greedy sources when there are a large number of greedy flows.

---

[17] The performance studies were done by S. Abraham and M. Chuah when they were at Bell Laboratories.

Next, we consider the round-robin service discipline. Table 8-3 tabulates the results. There is more than 50% increase in the throughput for the HTTP sources when compared to the FCFS case. The overall buffer drop probability also increases since HTTP flows can now send more traffic.

Table 8-2 Mean/Std Throughput for HTTP/Greedy Flows with FCFS Service Discipline

Num. of HTTP Flows	Num. of Greedy Flows	Avg. Thpt. of HTTP Flows in Active Period Kb/s	Std. Dev. of Thpt. of HTTP Flows	Tot. Thpt. of HTTP Flows Mb/s	Avg. Thpt. of Greedy Flows Kb/s	Std. Dev. Thpt. of Greedy Flows	Tot. Thpt. of Greedy Flows Mb/s	Buffer Drop Prob. (%)	Tot. Thpt. Mb/s
500	1500	21.9	2.66	3.9	26.9	7.556	40.3	2.23	44.2
700	1300	22.6	3.03	5.58	29.9	8.35	38.8	1.54	44.4
1000	1000	24.6	2.99	8.24	36.3	6.6	36.3	0.55	44.6
1300	700	38.1	1.52	12.15	43.8	0.233	30.7	0.06	42.8
1500	500	38.3	1.45	14.02	43.8	0.203	21.9	0.005	35.9

Table 8-3 Mean/Std Throughput for HTTP/Greedy Flows with Round-Robin Service Discipline

# of HTTP Flows	# of Greedy Flows	Avg. Thpt. of HTTP Flows in Active Period Kb/s	Std. dev. of HTTP Thpt.	Tot Thpt. of HTTP Flows Mb/s	Avg. Thpt of Greedy Flows Kb/s	Std. Dev. of Thpt. of Greedy Flows	Tot. Thpt. of Greedy Flows	Buffer Drop Prob. (%)	Tot. Thpt. Mb/s
500	1500	33.68	2.88	4.5	26.39	8.4	39.6	2.7	44.1
700	1300	35.09	2.77	6.39	29.11	9.59	37.8	2	44.2
1000	1000	36.1	2.33	9.24	34.48	10.81	34.5	0.9	43.7

To provide different QoS classes, we consider weighted round robin disciplines. We assume that HTTP users are assigned two different QoS classes. One set of users (Class1) are given a weight of 4, the other (Class 2) are given a weight of 2. The Greedy sources fall under Class 3 with a weight of 1. A tail drop policy is used to manage the buffer space for all classes of traffic. The results are tabulated in Table 8-4.

The results in Table 8-4 indicate that weighted round robin does not provide much improvement over the round robin case. The additional weight provided to Class 1 cannot overcome the effects of the tail drop policy. We repeat the same experiment using a buffer space of 25 Mbytes but similar results were obtained. So, we next investigate if any noticeable differences can be seen, by using a buffer management scheme that treats traffic from different classes differently. One such buffer management scheme is the weighted red scheme.

Table 8-4. Mean/Std. Thoughput for HTTP/Greedy Sources Using Weighted Round Robin Service Discipline

Buffer Size	Type/Weight	No. of sessions	Mean Thrpt. Kb/s	Standard Deviation	Total Thrpt. Mb/s	Buffer Drop Probability %
	HTTP 4	250	35.43	2.62		
2000	HTTP 2	250	33.28	2.53		2.8
	Greedy 1	1500	26.39	8.71	39.58	
	HTTP 4	350	37.69	1.54		
2000	HTTP 2	350	33.63	2.53		2.1
	Greedy 1	1300	29.08	9.83	39.58	
	HTTP 4	500	38.53	1.52		
2000	HTTP 2	500	38.34	2.56		2.1
	Greedy 1	1000	35.29	10.76	34.68	

For our experiments, we tried two sets of WRED thresholds:

1. HTTP Class 1: $P_{min} = 1800$, $P_{max} = 2000$
    HTTP Class 2: $P_{min} = 1700$, $P_{max} = 1800$
    Greedy: $P_{min} = 10$, $P_{max} = 1700$
2. HTTP Class 1: $P_{min} = 1800$, $P_{max} = 2000$
    HTTP Class 2: $P_{min} = 1700$, $P_{max} = 1800$
    Greedy: $P_{min} = 1450$, $P_{max} = 1700$

Traffic from different classes are served using the weighted round robin scheme with their weights as stated above, i.e., a weight of 4 for HTTP Class 1, a weight of 2 for HTTP Class 2, and a weight of 1 for Greedy sources. The results are tabulated in Table 8-5 (with Set 1 parameters), Table 8-6 (with Set 2 parameters).

Table 8-5 Mean/Std. Throughput for HTTP/Greedy Classes (Set 1 Parameters)

Buffer Size	Type/Weight	No. of sessions	Mean Thrpt. Kb/s	Standard Deviation	Total Thrpt. Mb/s	Buffer Drop Probability %
	HTTP 4	250	38.69	1.39	2.34	0.00
2000	HTTP 2	250	38.67	1.25	2.31	0.00
	Greedy 1	1500	24.47	6.43	36.71	3.20
	HTTP 4	350	38.75	1.49	3.32	0.00
2000	HTTP 2	350	38.68	1.28	3.28	0.00
	Greedy 1	1300	24.77	6.68	32.20	2.90
	HTTP 4	500	38.62	1.55	4.70	0.00
2000	HTTP 2	500	38.61	1.55	4.75	0.00
	Greedy 1	1000	25.42	7.07	25.42	2.50

Table 8-6. Mean/Std Throughput for HTTP/Greedy Classes (Set 2 Parameters)

Buffer Size	Type/Weight	No. of sessions	Mean Thrpt. Kb/s	Standard Deviation	Total Thrpt. Mb/s	Buffer Drop Probability %
2000	HTTP 4	250	38.70	1.37	2.35	0.00
	HTTP 2	250	38.64	1.53	2.33	0.00
	Greedy 1	1500	26.37	7.99	39.56	2.40
2000	HTTP 4	350	38.71	1.34	3.30	0.00
	HTTP 2	350	38.67	1.44	3.29	0.00
	Greedy 1	1300	29.07	9.09	37.79	1.80
2000	HTTP 4	500	38.77	1.33	4.72	0.00
	HTTP 2	500	38.65	1.41	4.74	0.00
	Greedy 1	1000	34.73	9.67	34.73	0.70

The results show that by giving different buffer space treatments to different QoS classes, the packet loss for HTTP classes can be eliminated and their throughputs also increase. Of course, this is at the expense of increasing the packet losses of greedy sources.

The above studies indicate that scheduling and buffer management schemes at GGSN play an important role in determining the end-to-end TCP throughputs of the packet data sessions in UMTS network. Careful design of both the scheduling and buffer management schemes needs to done before one can provide performance differences to the traffic from different QoS classes.

8.8 Distributed/Centralized Core Network Design

In a UMTS network, packet services are provided by transporting traffic between the radio access network and the UMTS core network. UMTS packet data traffic terminates at GGSNs. GGSNs extract packets and route them to and from the service provider's data centers and/or public data network point of presence (POIs). Clusters of SGSNs can be homed to regional GGSNs. The design issue is: How many clusters should there be and how should they be formed such that the total equipment and transport cost can be minimized?

A generic way to solve this problem is described in [Atk01]. In this paper, the authors attempt to get a qualitative sense of the tradeoffs that can be done via choosing different values for some key parameters. They consider a greenfield situation where there are M SGSNs with certain traffic demands uniformly distributed over a large area. Linear cost models for both the communication and equipment cost are assumed. The communication cost for a given bandwidth consists of a fixed base cost plus a per-mile charge per Mbps. Similarly, the infrastructure (equipment, power, space, etc) cost of a node consists of a fixed initial cost plus a cost per unit of capacity.

Let n be the number of GGSNs, C denote the fixed infrastructure cost per node, and B be the total distance-related bandwidth cost of the centralized architecture. The following expression is used for the variable costs of an architecture with n nodes.

$$\text{Variable Costs} = \frac{KB}{\sqrt{n}} + nC.$$

Normalizing by the centralized bandwidth B, the % reduction in bandwidth cost (net after including additional infrastructure cost), can be expressed as:

$$\% \text{ saving in variable cost} = 1 - (\frac{1}{\sqrt{n}} + n/(B/C))$$

Plots of the savings are shown in Figure 8-15 for several values of the ratio B/C. The plots indicate that a centralized approach is optimal only if B/C is less than 6.3 (approximated).

The big question is: Given n GGSNs and M demand sources (corresponding to SGSNs), where should the GGSNs be located so that the transport cost can be minimized? This is like the classical p-median location problem [Atk01]. In this combinatorial optimization problem, one is given a graph with leaves corresponding to user (demand sources), each node has a demand (possibly zero), edges have associated distances, and n nodes must be selected to minimize the average weighted distance from demand nodes to the nearest selected node. The "weight" corresponds to the demand. In our context, nodes with non-zero demand correspond to SGSN locations, the n selected nodes correspond to GGSN locations, and the edges correspond to backhaul network links. Solution of this p-median problem gives the minimum bandwidth location of n GGSNs. Overall optimization can be done by iteratively solving with $n = 1,2,3$ computing the total cost (bandwidth+infrastructure), and minimizing over n. Assuming that the overall cost function is convex, various methods can be used to avoid evaluating all values of n. In general, the p-median problem is computationally intractable (NP-hard); however, effective heuristics exist, particularly for modest-sized problems [Das95].

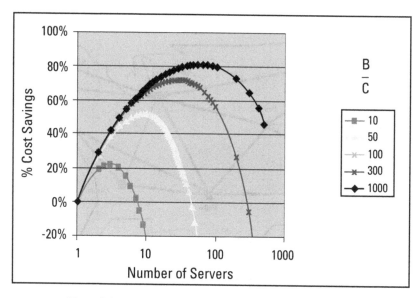

Figure 8-15 Centralized versus Distributed Tradeoffs: % Savings

The authors in [Atk01] use the Lagrangian relaxation method to solve the optimization problem for the network shown in Figure 8-16 assuming equal demand of 500 Mbps at all 10 nodes. Link costs were $0.54 per Mbps per mile per month (based on current OC-48 rates) and fixed node costs (GGSN) were assumed to be $15.5K per month ($500K amortized over 3 years at 7%). The above design was performed using a tool called UWIND that is developed by some Bell Laboratories researchers [UWIND]. Their results show that the minimum cost architecture involves placing GGSNs at all 10 locations with a total monthly cost of $155K as compared to a cost of $15.5 M with a centralized architecture (optimally locates at Chicago). Interested readers can refer to [Atk01] for more results assuming different facility types (DS-3, OC-3, OC-12, OC-48).

The above example indicates that a careful design of the UMTS system needs to be done to allow the operators to save monthly transport cost.

Acknowledgments
The results reported in Section 8.7 were from joint work done with former Bell Laboratories colleague Dr. S. Abraham while the results reported in Section 8.8 were from joint work done with Dr. G. Atkinson and Dr. S. Strickland. The results in Section 8.8 were published in a conference paper [Atk01].

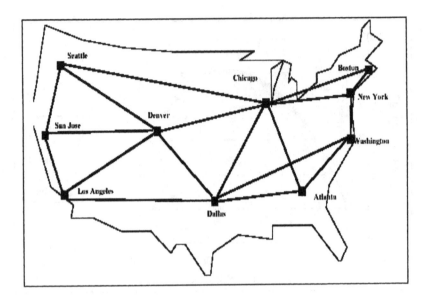

Figure 8-16 Backbone Network Example [Atk01]

The authors would like to thank 3GPP for giving permission to reproduce some of the diagrams from the 3GPP TSG standard documents. 3GPP TSs and TRs are the property of ARIB, ATIS, ETSI, CCSA, TTA, and TTC who jointly own the copyright for them. They are subject to further modifications and are therefore provided to you "as is" for information purposes only. Further use is strictly prohibited.

8.9   Review Exercises

1. How does 3G-SGSN differ from 2G-SGSN?
2. What are the functions of SGSN and GGSN? Design a system in which the functions of SGSN and GGSN can be merged into one logical entity. List the advantages and disadvanges of having only one special router in a UMTS Core Network rather than having two.
3. Develop a spreadsheet using the methodology described in Section 8.4.1 and Table 8.1 to verify the estimated signaling load numbers given in Section 8.4.1.
4. If the RAB idle time is set to 15 s, determine the total downlink/uplink signaling traffic generated by all Web-browsing users for PDP context activation/deactivation and RAB setup/teardown procedures across the Iu Interface. Hint: Assume that

there are 20 pages during one Web session and then the page reading time exceeds the RAB idle time so each new page requires a new RAB setup.

5. Estimate the total downlink/uplink signaling traffic generated by all Web-browsing users for PDP context activation/deactivation and RAB setup/teardown procedures across the Iu interface if all Web users are moving on the average 10 km/hr rather than 60 km/hr.

8.10 References

[23.060] 3GPP GPRS: Service Description: Stage 2, TS23.060 V3.15.0 (2003-06).

[29.060] 3GPP GPRS Tunnelling Protocol across the Gn and Gp interface, TS29.060 V3.19.0 (2004-03).

[Atk01] G. Atkinson, S. Strickland, M. Chuah, "UMTS Packet Service Infrastructure: Distributed Is Cheaper," Networks 2002, June 2002.

[Cha99] H. Chaskar, U. Madhow, "Fair Scheduling with Tunable Latency: A round robin approach," Proceedings of Globecom, 1999.

[Das95] M. Daskin, "Network and Discrete Location: Models Algorithms, and Applications," Wiley, 1995.

[Flo95] S. Floyd, V. Jacobson, "Link-Sharing and Resource Management Models for Packet Networks," IEEE/ACM transaction on networking, Vol 3, Nov 4, pg 365–386, August 1995.

[Ka01] S. Kasera et al, "Fast and Robust Signaling Overload Control," ICNP, November 2001.

[NS] NS Simulator: http://www-nrg.ee.lbl.gov/ns/

[RFC2661] W. Townsley etc, "Layer Two Tunneling Protocol," Internet Engineering Task Force RFC2661, August 1999.

[WRED]WeightedRed:http://www.cisco.com/univercd/cc/td/doc/products/software/ios112/ios112p/gsr/wred_gs.htm

Chapter 9

END-TO-END PERFORMANCE IN 3G NETWORKS

9. INTRODUCTION

Third-generation wireless systems was designed with the goal to provide multimedia services, e.g., voice, data, and video. Thus, it is of interest to find out how well the design meets its goal by studying the end-to-end performance of voice and data services in 3G networks. In this chapter, we first describe a methodology for estimating the call setup time for voice service, and the end-to-end bearer delay for voice services in Section 9.1. For packet data service, one important performance metric is the end-to-end Transport Control Protocol (TCP) throughput one can get from a 3G data connection since most of the wireless data applications are built on top of TCP/IP protocols. In order to understand TCP performance in 3G networks, one needs to understand first the radio link control (RLC) protocol that is designed to perform local link layer retransmissions to camouflage the higher frame error rate of wireless links. Thus, using UMTS as an example, in Section 9.2, we describe how UMTS RLC/MAC works. Then, we describe how one can study the RLC performance analytically and via simulations in Sections 9.2.1 to 9.2.3. Next, in Section 9.2.4, we discuss how RLC deadlocks can be avoided. In Section 9.3, we discuss how typical TCP parameters should be set to obtain good performance in 3G networks. Last but not least, we discuss various techniques that some researchers suggested for optimizing the TCP performance in 3G networks in Section 9.4.

9.1   Call Setup Delay for Circuit Switched Service

Figure 9-1 gives a high-level overview of the end-to-end path of a circuit voice call in UMTS network. Voice packets from an external source, e.g., a mobile in another wireless network or the PSTN destined for a mobile of interest are sent to its associated MSC. The MSC performs the switching

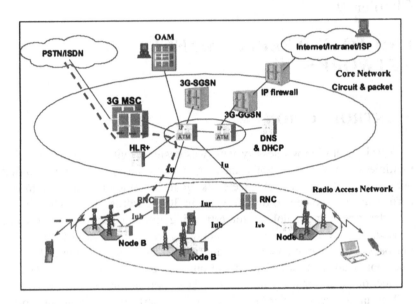

Figure 9-1. End-to-End Path for a Circuit Voice Call

function for the UMTS network and forwards the voice packets to the SRNC to be switched to the destined mobile. The MSC also performs functions such as authentication, admission control, and the collection of call details record for the billing server.

RNC is responsible for the framing and MAC functionalities, e.g., RNC performs MAC layer scheduling, power control function, etc. NodeB houses the transmission equipment and is responsible for delivering the packets over the airlink to the final destination. Information about the airlink, e.g., frame error rate is collected at NodeB and forwarded to the RNC.

We present an analysis of the call setup delay in a UMTS network for two scenarios: a Mobile-to-Mobile call and a Mobile-to-PSTN call. Here, we only consider mobile originated calls.

The following steps are taken to set up a mobile originated call:
1. A signaling channel is set up between the mobile and its associated MSC. This channel is established using the RRC connection establishment procedure. After the RRC connection is established, the mobile sends a service request message to the MSC. This message is carried to the MSC via the RNC using the RANAP protocol. The MSC responds to this service request by performing

an identification and authentication procedure to determine if the mobile is permitted to use the service.

2. Once it is determined that the mobile is authorized to use the network, the mobile sends a setup message containing the called party's number to the MSC. The MSC responds to this setup request by sending a set of signaling messages to RNC which initiates the procedures for setting up the transport channels for user plane data in the terrestrial network and Radio Access Bearer (RAB) assignment at the RNC.

3. For a Mobile-to-Mobile call, the MSC, which the destination mobile associates with, will initiate the paging procedure to contact the called mobile. Once the called mobile responds to the paging procedure, the MSC performs the identification and authentication procedures to determine if the called mobile is allowed to use the network. If the identification authentication is successful, a Radio Access Bearer (RAB) will be set up in the transport network. In addition, the airlink is established towards the called mobile.

4. If the call is destined to a PSTN phone, the MSC selects a trunk access gateway for the call and forwards setup messages to the PSTN using the ISUP signaling protocol. Once the call setup procedure is completed in the PSTN, a setup complete message is sent to the MSC.

5. On receiving the confirmation that the connection setup procedure on the called party's side is successfully completed, a ringing tone is sent to the calling party and the conversation phase can begin.

For the call release initiated by the calling user, a disconnect message is sent to the MSC which initiates teardown procedures for the terresterial transport channel and the RAB at both the calling side and the called side. The timer for the call is stopped and the call duration information is then sent to the billing server.

9.1.1 Delay Analysis of the Call Setup Procedure[18]

Figure 9-2 shows a detailed message flow for establishing a 3G-to-3G call. For a UMTS Mobile-to-Mobile call, there are seven major steps that a calling mobile takes to make a call. Step 1 is the RRC setup procedure [25.331]. Steps 2,3, and 4 deal with the identification and authentication of

---

[18] This analysis was done by Dr. S. Abraham while he was at Bell Laboratories. The authors thank Lucent Technologies for allowing us to use the data.

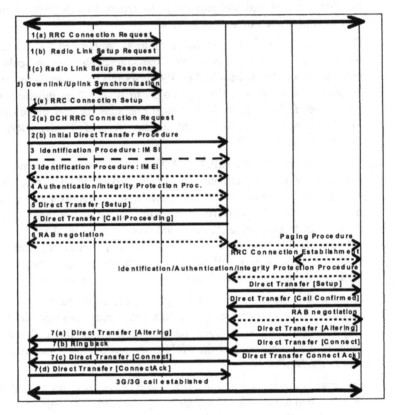

Figure 9-2. Message Flow for Establishing a 3G-to-3G Call

the calling mobile. In Step 5, called party information is sent to the Call Server. In Step 6, the Radio Access Bearer (RAB) is set up for the transfer of user data. This step also involves the creation of the Iu and Iub interfaces for the call. Once the procedures associated with the called mobile are performed, an alert message and ringing tone is sent to the calling mobile (Step 7). For the called mobile, the first step is for the network to page for the mobile to inform it of the incoming call. Subsequent procedures taken by the called mobile is similar to the calling mobile.

The estimated delay for each procedure taken by the calling side is summarized in Table 9-1 while the estimated delay for each procedure taken by the called mobile is summarized in Table 9-2. The estimates have been obtained using estimates of the required messages sizes and the processing times from some laboratory studies. Educated guesses have been used for some components of the delay for which information was not available.

Table 9-1. Estimated Delay for Call Setup Procedures for the Called Mobile

	Components Call Setup Procedure from Calling UE to 3G MSC	Delay (ms)
1.	RRC connection establishment	264.5
2.	NAS signaling connection establishment	85.3
3.	NAS authentication procedure	2005.3
4.	Integrity protection procedure	244.7
5.	NAS call setup procedure	566.2
6.	Synchronized RAB assignment procedure – circuit switched	512.5
7.	NAS alerting/connect procedure – mobile originated	467.0
	Subtotal delay from calling mobile to 3G MSC	4145.5

Table 9-2. Estimated delay for Each Procedure Taken by the Calling Side Mobile in a Mobile-to-Mobile Call

	Call setup procedures from called mobile to 3G MSC (assume the called UE served is by the same MSC)	Delay (ms)
1.	Paging	41.7
2.	RRC connection establishment – DCH establishment	264.5
3.	NAS signaling connection established	85.3
4.	NAS authentication procedure	205.3
5.	Integrity protection procedure	244.7
6.	NAS call setup procedure	2366.2
7.	Synchronized RAB assignment procedure – circuit switched	512.5
8.	NAS alerting/connect procedure – mobile terminated	468.9
	Subtotal delay from called mobile to 3G MSC	4189.1

The total call setup delay is obtained by adding the delays incurred at both the calling and the called mobiles. The total delay from the calling side is about 4.19 s while the total delay for the called side is about 4.15 s. So, the total mobile-to-mobile call set up delay can be as long as 8.34 s.

9.1.2 End-to-End Delay Analysis for Voice Bearer

In this subsection, we identify the main components of the end-to-end transport delay for voice calls. We also identify the excess delay introduced by the wireless access portion of the network as compared to the baseline delay from an equivalent wireline solution. For wireline networks, transport delay budgets are typically identical for both directions of the user data flows. For wireless networks, the network access mechanism differs in the downlink and uplink directions. In addition, different styles of bearer services for voice traffic (e.g., circuit-switched or packet-switched) may be available both in the access and core networks.

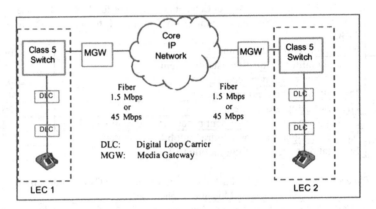

Figure 9-3. Reference Model for VoIP in the Core Network [Her00] © IEEE Copyright 2000

Figure 9-3 [Her00] shows a reference model for voice traffic over a wireline network assuming an IP-capable core network. Here, only the core portion of the transport network is assumed to be IP-based, while the access portion uses the existing PSTN. Since the access portion is PSTN-based, user data flows are symmetric. This scenario is presented as a reference scenario.

Five basic subsystems are illustrated in Figure 9-3:
- Ingress Local Exchange Carrier (LEC) Network
- Ingress Media Gateway
- Core IP Network
- Egress Media Gateway
- Egress LEC Network

The media gateway(MGW) is responsible for media interworking functions including speech packetization, streaming, and playout between the circuit-based and packet-based portions of the voice service network. Table 9-3 shows a delay budget for this reference wireline network.

Delay objectives for the PSTN portion of the wireline network are derived from existing standards, e.g., [ETR102], [G.114], or laboratory experiments, e.g., [Luc99]. In addition, we assume identical contributions from both the ingress and egress LEC portions of the voice service network. The delay objective for the LEC assumes a network span of 250~500 miles and includes processing delay contributions from access network switches and the terminating Digital Loop Carrier (DLC) equipment. The delay objectives for the core network assumes either a short span (2000 miles) high-capacity broadband network (OC-3c rates and above) or a long-span

(4000 miles) broadband network with sub-broadband access links (a few DS-1/DS-3 links). The delay objective for the Ingress MGW assumes 20 ms PCM voice frames per IP packet. Note, however, that there are engineering and operational tradeoff associated with the chosen voice frame size, the number of voice frames per IP packet and the transport network operational capacity. These tradeoffs are studied in greater detail in [Dos98]. The delay objectives for the Egress MGW assume either a dedicated and well-engineered core network (with low transport delay variation) or a shared and more loosely engineered core network (where a more conservative playout strategy may be required). The shared network scenario is considered because a dedicated packet network for the voice traffic may be unrealistic for many practical deployment scenarios.

Table 9-3. Wireline One-Way Transfer Delay Across an IP-Based Core Network [Her00]

Subsystem	Delay Components	Nominal Values	Remarks
Ingress LEC	Tiloop	4.5	Ingress loop processing
	Tprop-co	2.0–4.0	Propagation delay to CO switch
	Tswt	1.5	Class 5 switch processing
	Tprop-gw	1.2	Propagation delay to MGW
		Total I-LEC = 9.2–11.2	
Ingress Media Gateway	Tsample	20	Packetization delay: 2x10ms PCM frames
	Tigw	0.5	Ingress MWG processing
	Tmargin	0–0.5	Other I-GW processing delay
		Total I-GW = 20.0–21.0	
Core IP Net	Tdcs	1.0–2.0	Processing delays in transmission network
	Trouting	1.0–7.0	Processinjg and queueing delays in all core routers
	Tprop-cn	16.0–32.0	Propagation delay across core network
		Total Core IP Net =18.0–41.0	
Egress Media Gateway	Tjitter	2.3–18.3	Jitter compensation delay: 2x10ms PCM frames
	Tegw	0.5	Egress MGW processing
	Tmargin	0.0–0.5	Other processing delay
		Total E-GW = 2.8–19.3	
Egress LEC	Tprop-gw	1.2	Propagation delay to MGW
	Tswt	1.5	Class 5 switch processing
	Tprop-co	2.0–4.0	Propagation delay to CO switch
	Teloop	4.5	Egress loop processing
		Total E-LEC = 9.2–11.2	
		Total = 59.7–103.7	

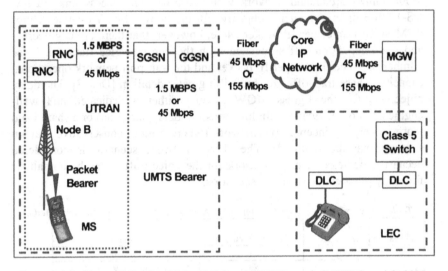

Figure 9-4. Reference Model for a Voice Call Over UMTS Network © IEEE Copyright 2000

Figure 9-4 shows the reference model for a voice call between a mobile over the UMTS network and a regular PSTN phone. Eight basic subsystems are involved:

- LEC Network
- Media Gateway
- Core IP Network
- GGSN
- SGSN
- RNC
- NodeB
- Mobile

Figure 9-4 shows the UMTS reference model when the connection spans across two RNCs with the Iur interface connecting the two RNCs in a soft-handover scenario. Note that a soft-handover case may not span across two RNCs. The mobile may talk to two NodeBs that are connected to the same RNC. In that case, only one RNC is involved and that RNC talks directly to the SGSN.

There are two major scenarios of interest. Either the MS generates VoIP traffic directly or it generates AMR voice frames to be later transported over an IP-capable wireline network. For the second scenario, we assume that the

AMR voice packets are encapsulated and multiplexed within IP packets by the RNC. Here, we describe how the end-to-end voice delay can be determined. In our voice delay analysis, we consider an equivalent configuration to the wireline scenario where the UTRAN access network replaces one of the LEC networks. Thus, the same delay objectives apply to the shared portions of the transport network, namely, Ingress LEC, Ingress MGW, and core IP-based network. Delay objectives for the UMTS radio bearer are derived from 3GPP contributions [3Gdelay], [23.060], and Bell Laboratories performance studies.

Downlink Delay Analysis

Table 9-4 summarizes the delay contributions on the downlink direction of the user flow. The GGSN supports all the functions typically associated with the wireline Egress GW. Yet, there is no jitter compensation performed in the GGSN. This function resides in the MS. The SGSN also supports similar packet relay functions as the GGSN. Additional control procedures are performed at SGSN, which incur additional processing resources. The RNC supports more than just packet relay functions. Processing requirements for the RNC are expected to be higher than in the SGSN/GGSN. Extra processing resources need to be allocated to support RF processing, soft-handover procedures as well as any ATM portion of the wireline network. Transcoding is supported at the Ingress Gateway.

Uplink Delay Analysis

Table 9-5 summarizes the delay components in the uplink direction of the user flow. The RNC, SGSN support relay functions similar to those of the GGSN. The RNC also supports uplink soft-handover procedures. Additional processing resources, e.g., for handling session management related functions may be required for RNC and SGSN since they are the access points into the packet and backhaul networks. NodeB needs resources for RF processing as well as soft-handover procedures. Egress MGW performs uplink transcoding from AMR to PCM. In addition, all uplink jitter compensation for the delays introduced by the packet network are supported by the Egress MGW.

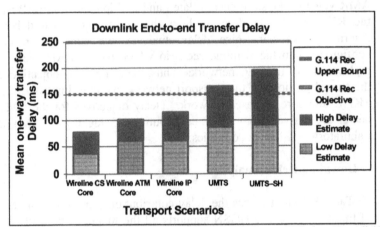

Figure 9-5. One-Way Transfer Delay for Reference Wireline and Wireless (Downlink) Configurations © IEEE Copyright 2000

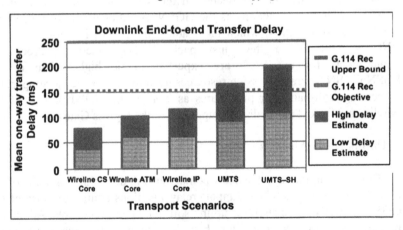

Figure 9-6. One-Way Transfer Delay for Reference Wireline and Wireless (Uplink) Configurations © IEEE Copyright 2000

Figures 9-5 and 9-6 show the end-to-end transfer delay for packet voice over UMTS (downlink and uplink, respectively) when compared with wireline based solutions. The label SH in the figures refers to soft-handover scenarios. One-way transfer delay for our reference wireline scenario is substantially lower than the recommended objective in ITU-T G.114 since we assume relatively shorter network spans and digital facilities. For wireless solutions, even under less than conservative processing assumptions, its one-way delay tends to be closer to the upper range of the recommended delay objective. A substantial fraction of the additional delay is due to packetization, de-jittering, and RF processing functions specific to the wireless access network rather than specific to the packet-based transport mode. Note that the excess delay introduced by the soft-handover procedures is relatively small compared to the other contributors to the overall end-to-end delay. It should be pointed out that we assume that broadband links are used to interconnect the various ATM switches. Transfer delays and jitters may be larger if non-broadband links are used.

Table 9-4. Packet Bearer Downlink Transfer Delay

Sub-system	Delay Component	No SH	With SH	Remarks
ILEC		9.00–11.20	9.00–11.20	From Bell Laboratories Internal Studies
Ingress GW	Tsample	20.00	20.00	Assumes 20 ms PCM speech sample per packet
	Ttransc	4.00	4.00	Transcoding processing to 12.2 kbps
	Tigw	0.50	0.50	I-GW switching
	Tmargin	0.0–0.5	0.0–0.5	I-GW other processing
	Total I-GW delay=24.5–25.		24.5–25.0	
Core IP Net		18.0–41.0	18.0–41.0	From Bell Laboratories Internal Studies
GGSN	Tjitter	0.0	0.0	No de-jittering required
	Tggsn	0.5	0.5	GGSN switching
	Tmargin	0.0–0.5	0.0–0.5	GGSN other processing
	Tfiber	0.5–1.0	0.5–1.0	100 – 200 Km propagation delay
	Total GGSN delay= 1.0–1.5		1.0–1.50	
SGSN	Tdsgsn	0.5–1.5	0.5–1.5	SGSN processing
	Tmargin	0.0–0.5	0.0–0.5	SGSN other processing
	TN1 – Iu	0.5–1.0	0.5–1.0	ATM packetization and switching
	TN2 – Iu	0.5–1.0	0.5–1.0	100 – 200 km Media delay @ 5µs/km
	Total SGSN delay= 1.5–4.0		1.5–4.0	
RNC	Tsrnc	1.0–5.0	1.0–5.0	Processing delay in the SRNC
	Tmargin	0.0–0.5	0.0–0.5	RNC other processing
	Tn1 – Iub	0.6–1.2	0.6–1.2	ATM packetization and queueing delay (3 ATM cells)
	TN3 – Iub	0.6–1.2	0.6–1.2	Iub ATM switching delay
	TN2 – Iub	1.0–2.0	1.0–2.0	200 – 400 km @ 5µs/km
	TN1 – Iur		0.6–1.2	ATM packetization and queueing delay
	TN2 – Iur		0.5–3.0	100 – 600 km Media delay @ 5µs/km
	TN3 – Iur		1.20	ATM Switching delay
	TDRNC		1.0–5.0	DRNC Processing delay
	Total RNC delay = 3.2–9.9		6.5–20.3	
NodeB	Tbuff-align	0.0–10.0	0.0–10.0	Alignment with transcoder frame – may be zero
	Tbuff-atm	1.0–9.8	1.9–18.8	Delay jitter due to ATM connection

Table 9-4. Cont.

Sub-system	Delay Component	No SH	With SH	Remarks
NodeB	Tencode	1.0	1.0	Time alignment procedure for inbound control of the remove transcoder interface in the downlink direction
	Tmargin	0.0–1.1	0.0–1.1	Processing delay variation due to queueing, packetization
	Tsho		1.0	Soft handover delay
	Tot. Node B delay =2.0–21.9		3.9–31.9	
MS	Trftx	20.0	20.0	Interleaving and de-interleaving
	Trxproc	4.0	4.0	Equalization and channel decoding
	Tproc	2.0	2.0	Time required after reception of the first RP sample to process the speech encoded data for the full-rate speech decoder and to produce the first PCM output sample
	Tprop	0.1	0.1	Airlink propagation delay
	Tmargin	0.0–1.0	0.0–1.0	MS other processing delay
	Tplayout	4.3–30.3	4.3–30.3	Playout buffer delay (jitter compensation)
	Total MS delay =30.4–57.4		30.4–57.4	
	Total delay =89.6–171.4		94.8–192.3	

Table 9-5. Packet Bearer Uplink Transfer Delay

Subsystem	Delay Components	No SH	With SH	Remarks
Egress LEC		9.2–11.2	9.2–11.2	From Table 9-4
Egress MGW	Tplayout	4.3–37.1	5.2–46.1	Playout buffer delay (jitter compensation)
	Tproc	0.8	0.8	Transcoding processing to PCM
	Tegw	0.5	0.5	E-GW switching delay
	Tmargin	0.0–0.5	0.0–0.5	I-GW other processing delay
	I-GW delay = 5.6–38.9		6.5–47.9	
Core IP Net		18.0–41.0	18.0–41.0	From Table 9.4
GGSN	GGSN delay = 1.0–1.5		1.0–1.5	Initial allocated budget
SGSN	SGSN delay = 1.5–4.0		1.5–4.0	Initial allocated budget
RNC	Tdnrc	1.0–5.0	1.0–5.0	DRNC processing delay
	Tmargin	0.0 – 0.5	0.0–0.5	RNC other processing delay
	TN3-Iub	0.6 – 1.2	0.6–1.2	Iub ATM switching delay
	TN2-Iub	1.0 – 2.0	1.0–2.0	200-400 km @ 5 μs/km
	TN1-Iur		0.6–1.2	ATM packetization & queueing delay
	TN2-Iur		0.5–3.0	100-600 km Media delay @ 5 μs/km
	TN3-Iur		0.6–1.2	ATM switching delay
	Thbuff		10.4	Frame selection jitter delay
	Tsrnc		1.5–5.0	SRNC processing delay
	RNC delay =2.6–8.7		15.7–29.5	

Table 9-5. Cont.

Subsystem	Delay Components	No SH	With SH	Remarks
Node B	TN1-Iub	0.6–1.20	0.6–1.2	ATM/AAL2 packetization delay. Max is 10 ms
	Trxproc	4.0	4.0	Equalization and channel decoding.
	Tmargin	0.0–2.0	0.0–2.0	Other processing delay
	Trftx	20.0	20.0	Assumes alignment over 20 ms
	Tsho	0.0	1.0	Softer hand-over delay
	Node B delay=24.6–27.2		25.6–28.2	
MS	Tencode	1.50	1.50	Time required for the channel encoder to perform encoding
	Ttransc	8.0	8.0	Speech encoder processing time
	Tsample	25.0	25.0	20 ms frame + 5 ms from next frame
	Talign	1.0	1.0	Alignment of speech and radio frame
	Tprop	0.1	0.1	RF link propagation
	Tmargin	0.0–0.5	0.0–0.5	Other processing delay
	Techo	2.0	2.0	Zero in GSM
	MS delay = 37.6–38.1		37.60–38.1	
	Total delay =100.1–170.6		115.1–201.4	

9.2 End-to-End TCP Throughput in 3G Networks

Most of the data applications are built on top of TCP since TCP provides end-to-end reliability via retransmissions of missing IP packets. TCP was originally designed for wired networks where the packet losses are due to network congestion and hence the window size of TCP is adjusted upon detection of packet losses. However, the packet losses in wireless networks are due mostly to bad radio conditions and not to network congestion. Errors in the airlink are caused by several factors e.g., interference from other sources, fading due to mobility, and scattering due to a large number of reflecting surfaces. Even though Cyclic Redundancy Correction (CRC)-based error correcting codes are attached to radio frames, they are not sufficient to provide the desired reliability. To provide the increased reliability, acknowledgments and retransmission schemes are designed in the radio link control protocols in wireless networks. In both CDMA2000 and UMTS networks, the designed radio link control protocols are Negative Acknowledgment (NAK)-based protocols. When both the transport layer and the MAC layer try to recover lost packets simultaneously, the end-to-end performance may be worse than when only one layer tries to recover the lost packets. Thus, it is important to understand how RLC/MAC parameters affect end-to-end TCP performance in 3G networks.

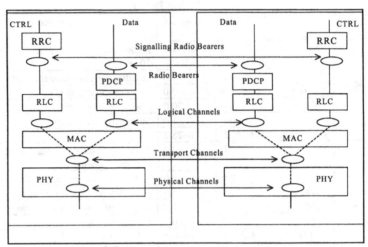

Figure 9-7. Simplified Radio Interface Architecture

We first describe the RLC mechanisms provided in UMTS standards. The UMTS RLC layer [25.322] is designed to provide reliable transport for packets from higher layers through the airlink. Figure 9-7 illustrates a simplified radio interface architecture which shows that the RLC layer accepts both control plane and user plane packets and transfers them to the logical channels. The receiving RLC layer confirms and delivers to higher layers RLC Service Data Units (SDUs) that have been received. The receiving RLC also indicates to higher layers which RLC SDUs have been discarded.

The MAC layer takes packets from the logical channels and places them in transport channels. The MAC layer selects the appropriate Transport Format Combination (TFC) for each transport channel depending on the user allocated rate. The MAC layer then informs the RLC the number of PDUs per transmission time interval (TTI). The MAC layer also schedules and coordinates the transmission of each RLC entity on the common and shared channels.

In Figure 9-8, we illustrate how a TCP/IP packet generated in the application layer is transformed into MAC Protocol Data Units (PDUs). The TCP/IP packet is passed to the Packet Data Convergence Protocol (PDCP) layer where a PDCP header will be inserted. The combined PDCP header and PDCP PDU constitute the RLC SDU. The RLC SDU is segmented into multiple RLC PDUs. The size of each RLC PDU is dependent on the

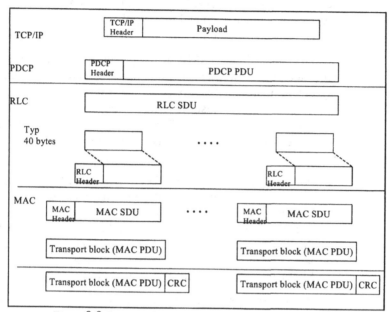

Figure 9-8. Segmentation of TCP Packets into RLC/MAC PDUs

allocated airlink rate, and the transmission interval (TTI) set for the user. The RLC PDU is 40 bytes if the TTI is 20 ms and the allocated airlink rate is 64 Kbps. A RLC header is inserted to each RLC PDU if necessary. The combined RLC header and the RLC PDU constitute the MAC SDU. MAC header will be inserted to each MAC SDU and the resulting packet is called a transport block or MAC PDU. The transport block is passed to the physical layer where Cyclic Redundant Code (CRC) is added. These transport blocks will undergo encoding, interleaving, rate matching, and multiplexing in the physical layer before they are transmitted on the airlink.

There are three modes for UMTS RLC:

(a) Transparent Mode (TM): Data transmission through the transparent mode is carried out without adding any overhead. When packets are received with error in the transparent mode, they may be dropped or forwarded to the higher layer. This mode is suited for the transmission of streaming applications.

(b) Unacknowledged Mode (UM): The difference between UM and TM mode is that sequence numbers are provided in the UM. A header is added to higher layer data to carry sequence number information. A timer based discard mechanism is available at the sender; however no discard information is communicated to the receiver. A segmentation and reassembly mechanism is provided. A candidate application for this mode of transmission is VoIP.

(c)   Acknowledged Mode (AM): In the presence of high airlink bit error rates, the acknowledged mode can be used to provide higher layers with more reliable service. In the AM, an automatic repeat request (ARQ) mechanism is used for retransmission of packets that are received with error. The receiver sends status messages to the sender to indicate the last packet received and the lost packets. Two feedback methods are provided:

   i.   Polling: The header of the RLC PDU has a bit that can be turned on to indicate that the receiver must send a status message about the currently received packets and missing packets. There are eight different polling triggers defined in the RLC specifications, e.g., last PDU in transmission buffer, last PDU in retransmission buffer, poll timer, poll every 'n' SDUs, window based, timer based, polling inhibit, etc.

   Polling triggers control the transmission of polling requests. Polling requests should be sent frequently so that the acknowledgments through status reports of the RLC transmission can be received quickly by the sender. The transmitter can either retransmit the lost PDU or remove the PDUs from the retransmission buffer (because the PDUs have been retransmitted a maximum number of times or have stayed in the retransmission buffer for too long) and the transmit window is moved to prevent stalling. The "last PDU in transmission buffer" and the "last PDU in retransmission buffer" polling triggers are provided to prevent deadlock of each RLC entity.

   ii. Receiver driven triggered: The receiving RLC can be configured to send status message periodically to the sender. The receiving RLC can also be configured to send status messages when missing PDUs are detected.

Upon receipt of a poll request or when periodic timer expires, the receiver always sends back a status report that is carried by one or multiple status PDUs.

At the sender, higher layer packets that have not been received at the receiver may be discarded when one of the following conditions holds:
   • The packet has been retransmitted a maximum number of times,

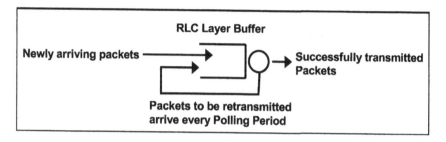

Figure 9-9. Model of the RLC Analysis

- The packet has been in the queue for more than a certain time.

The different polling triggers defined in the standards provide the flexibility and allow the network to control the polling frequency based on the network conditions, such as traffic load, link quality, quality of service requirements, etc. The choice of the polling trigger, discard strategy, and other associated parameters, e.g., polling period, have significant impacts on the performance of the RLC layer. The delay performance will improve with fast polling request but the more frequent polling requests result in more status reports and hence increase the overhead in the reverse link.

9.2.1   Simple Analytical Model for Studying RLC Performance

We present a simple analytical model[19] for studying the RLC performance assuming that the periodic status report feature is turned on at the receiver. A more sophisticated model can be found in [Qin02] which will be discussed in the next subsection.

In Figure 9-9, a schematic of the analytical model we consider is given. A status message is received every polling interval indicating which packets need to be retransmitted. The packets to be retransmitted are placed ahead of the packets that have not yet been transmitted. In order to construct a Markov chain, we observe the queue at the polling epochs. The arrival is modulated by a two-state on/off Markov chain whose transitions occur at the polling epochs. The number of packets sent in a polling period during the on state is fixed. No packets are sent in the off state. We assume that status messages are received without errors. Under the above assumptions, the joint process (queue length, state of the modulating Markov chain for the input

---

[19] This work was jointly done by Dr. S. Abraham and Dr. M. Chuah and documented in an unpublished memorandum while they were at Bell Laboratories.

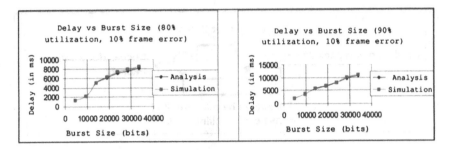

Figure 9-10. Airlink Utilization = 80%   *Figure 9-11.* Airlink Utilization = 90%

process) is a Markov process from which the queue length distribution of the queue can be obtained.

To derive some numerical values, we make the following assumptions:
- polling period is set to 8 frame transmission times with each frame time being 20 ms.
- Frame errors are assumed to be independent identically distributed with an average error rate of 10%.
- The airlink bandwidth is assumed to be 8 Kbps and the airlink utilization is assumed to be either 80% or 90%.
- Maximum number of frame arrivals in a polling period is assumed to be equal to the number of frames in the polling period.
- For each utilization value, the burstiness of the source process is varied and the distribution of the queue length process is obtained.
- The packet size is assumed to be 1000 bytes.
- Periodic packet arrivals are assumed. The packet interarrival times considered are 6.25, 25, 50, 100, 200 polling periods.

The analytical results using this model and the simulation results from NS simulation are presented in Figures 9-10 and 9-11. The figures show the 95% delay with different packet sizes using the analytical model and NS simulator match closely.

9.2.2   Analytical Model of RLC

Another analytical model for the periodic polling based retransmission scheme was developed to analyze the RLC performance [Qin02]. Figure 9-12 depicts the polling based retransmission procedure. In this model, we assume that the polling period is T_p s. T_s (s) is the transmission time of a PDU. RTT is the round trip propagation and processing delay between the transmitter and receiver. We assume that the probability of successful PDU

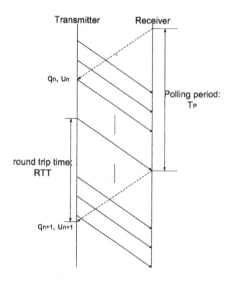

Figure 9-12 Model Polling Based Retransmission Procedure

transmission is p and the transmission errors are independent. At time $t = nT_p$, the transmitter receives the status report, removes all successfully transmitted PDUs from the transmission buffer, and places the negatively acknowledged PDU ahead of the buffer for transmission during the next polling interval.

We define the PDU delay, D, as the time that the PDU stays in the transmission buffer, i.e., the interval between the arrival time and the receipt of a positive acknowledgment to the PDU at the transmitter. In order to obtain the delay D, we first derive the queue length distribution. At time $t = (n+1)T_p$, the queue length is given by

$$q_{n+1} = q_n - s_n + a_n \tag{9-1}$$

where a_n is the total number of new arrivals during the interval $nT_p \sim (n+1)T_p$; s_n is the number of PDUs with positive acknowledgment. We assume that the PDU arrival follows Poisson distribution with a mean arrival rate λ. Therefore a_n is Poisson distributed random variable with mean equal to λT_p.

For simplicity of the analysis, we adopt a gating assumption that all new PDUs can enter the system only at the end of the polling interval and become eligible for transmission. The gating system PDU delay $\tilde{D}$ can be used to bound the real system delay as

$$E[\tilde{D}] - T_p/2 < E[D] < E[\tilde{D}] + T_p/2 \qquad (9\text{-}2)$$

By using Little's law, the delay of the gating system is given by

$$E[\tilde{D}] = \overline{q}/\lambda \qquad (9\text{-}3)$$

where $\overline{q}$ is the average queue length of the gating system.

In (9-1), s_n follows a binomial distribution based on the independent transmission error assumption,

$$s_n \sim Bi\left\{ \min\left[q_n, u_n + \left\lfloor (T_p - RTT)/T_s \right\rfloor \right], p \right\} \qquad (9\text{-}4)$$

and u_n is the number of outstanding PDUs during interval $nT_p \sim (n+1)T_p$, given by

$$u_n = \min\left\{ RTT/T_s, \max\left[0, q_{n-1} - u_{n-1} - \left\lfloor (T_p - RTT)/T_s \right\rfloor \right] \right\} \qquad (9\text{-}5)$$

In general, equations (9-1), (9-4), and (9-5) define the evolution of the this polling based retransmission model and can be used to construct the Markov chain.

The task of obtaining closed-form solution for this Markov chain seems formidable. However we are able to obtain a closed-form solution for the average system occupancy for some special cases. The detailed analysis can be found in [Qin02]. Here we summarize the analytical results and compare them with simulation results.

The parameters used for simulation and analysis are: PDU size equals to 40 bytes, transmission rate is 64 Kbps, TTI is 20 ms. As shown in Figure 9-13, the upper bound becomes loose as the polling period increases. The lower bound is always close to the simulated results. Figure 9-14 shows the simulated PDU delay and the upper and lower bounds when the RTT is larger than zero. It is shown that the lower bound is tight when the RTT is small or approaching the polling period, but becomes loose for RTT around half of the polling period. With this observation, one can better approximate the PDU delay by

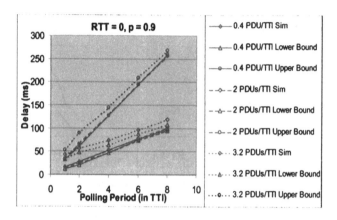

Figure 9-13 Simulated Queuing Delay and the Bounds for Different Arrival Rates © IEEE
Copyright 2002 [Qin02]

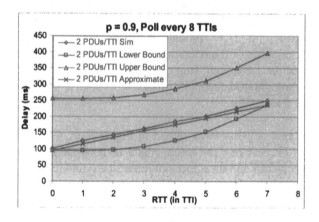

Figure 9-14 Simulated Queuing Delay and the Bounds with Non-zero RTT © IEEE
Copyright 2002 [Qin02]

$$\hat{D}_\tau = \hat{D}_0\left(1 - \frac{\tau}{T_p}\right) + \left(\hat{D}_0 + T_p\right)\frac{\tau}{T_p} = \hat{D}_0 + \tau \qquad (9\text{-}6)$$

It can be seen from the figure that this approximation is quite good.

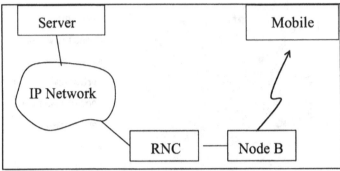

Figure 9-15. Simulated UMTS Network

9.2.3   Simulation Studies of RLC/MAC Performance

The above analysis assumes a packet generation process that is not controlled by any transport protocol. However, most Internet applications are built on top of the TCP protocol. Thus, it will be interesting to find out how the end-to-end TCP throughput varies with different RLC/MAC settings. The analysis of the operation of TCP protocols over the RLC layer is complicated, so subsequent studies of the impacts of RLC features on the end-to-end throughput/delay performance are carried out via simulations using the NS-2 simulator.

The simulated network is as shown in Figure 9-15. The server can be a FTP or a HTTP Server. The intermediate IP network is assumed to have sufficient bandwidth and thus does not present any bottlenecks to the traffic between the server and the mobile. The link between the RNC and NodeB is assumed to be of sufficiently high bandwidth, however, we assume that the delay between the two elements is fairly high. In the current experiments, we assume that transmission over the airlink is carried over a 64-Kbps dedicated channel with a TTI of 20 ms. The errors on the airlink are assumed to be independent identically distributed.

The parameter settings for the experiments are as follow:
- Periodic Polling messages are sent with a polling period of 200 ms.
- Airlink bandwidth of 64 Kbps and TTI of 20 ms.
- RLC PDU size is 40 bytes. Thus, 4 RLC PDUs can be sent in each frame for 64-Kbps service.
- TCP packet size of 1000 bytes.
- Airlink frame error rate is 10%.

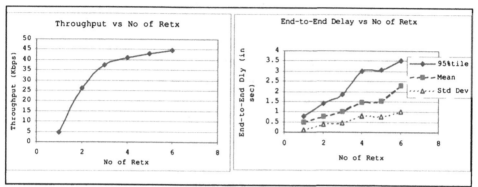

Figure 9-16. Throughput v. No. of Retx   *Figure 9-17.* E2E Delay v. No. of Retx

- The round trip delay between the RNC and NodeB is set to 100 ms. This delay excludes the queueing delay at the RNC.

To study the impact of the number of retransmissions, we vary the number of retransmissions from 0 to 6 and measure how the FTP throughput varies with the number of retransmissions. It is assumed that a RLC PDU will be dropped after it has been retransmitted a certain number of times. Figures 9-16 and 9-17 show how the TCP throughput and the end-to-end delay vary with the maximum number of retransmissions. For the end-to-end delay curve, we plot the mean, standard deviation and the 95th percentile delay. The plot in Figure 9-16 indicates that the end-to-end throughput almost doubles when the maximum number of retransmissions changes from two to five. The plot in Figure 9-17 indicates that such an increase in the end-to-end throughput is at the expense of increasing end-to-end packet delay due to retransmissions. Figure 9-18 shows how the RLC PDU transmission queue length varies with the maximum number of retransmissions. Again, this plot indicates that the increased throughput leads to a larger TCP window and hence an increasing buffer size requirement. Sufficient buffers need to be provided to avoid packet drops as a result of buffer overflows.

Packet dropping can cause huge reductions in the end-to-end TCP throughput, since TCP sessions wait for a "timeout" period before transmitting unacknowledged TCP packets. During these "timeout" periods, the allocated airlink capacity is unused resulting in poor throughput performance.

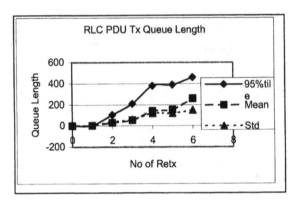

Figure 9-18. RLC PDU Transmission Queue Length v. No. of Retx

9.2.4 Deadlock Avoidance in RLC

The RLC protocol described in [25.322] contains many different options. If these options are not used appropriately, the RLC protocol can enter into a deadlock situation. For example, if we turn on only periodic polling, the protocol will stall under the following conditions:

1. The transmission buffer becomes empty before the end of a polling period. Some of the PDUs may have to wait till new SDUs arrive before a poll is sent.
2. The transmit window limit may have reached before the end of a polling period.
3. A lost status message prevents the window from advancing. The loss of the forward PDU with the polling bit set or the dropping of the status message can cause a loss in status messages.

The stalling condition 1 can be solved by sending a poll every time the transmission and retransmission buffers transition to the empty state. However, such a trigger cannot solve problems caused by conditions 2 and 3 described above. Two solutions are proposed to solve such stalling:

1. VT(S)-1 Approach [Su03]
A protocol stall is detected if any one or both of the following conditions occur:

- The transmit window limit has been reached and there are new PDUs available for transmissions.
- The transmit buffer is empty but there are previously transmitted PDUs waiting for acknowledgements.

When any or both of the above conditions are detected, a timer is started. The timer is stopped when a status report is received. Otherwise, the last transmitted PDU is retransmitted with a poll bit set when the timer expires. This is the VT(S)-1 approach described in [Su03].

2. Combined Poll-Window/Poll Timer Approach

Protocol stalling due to lost status messages (condition 3 discussed earlier) is prevented by using the poll timer feature. To prevent stalling that is caused by the end of the transmit window (condition 2), a window-based polling trigger is used.

A performance comparison of these two approaches was carried out via simulations.[20] In this simulation effort, the data application is assumed to be HTTP. The model for HTTP is taken from [Mah97]. This HTTP model assumes some uplink messages are sent to request for a page and additional uplink messages will be sent to download objects embedded within a page. After a page is downloaded, user think time will be incurred before another page is downloaded. The model parameters are as follow:

- Long-term mean page size = 35 Kbytes
- Mean number of objects per page = 5.5
- Think Time is assumed to be uniformly distributed between 5-45 s. Hence the mean think time = 25 s.

Since the HTTP traffic is bursty, we do not use periodic polling, instead we use "poll every *n* SDU" trigger. The maximum number of retransmissions is set to 5. We also compare the two methods described earlier that prevent RLC protocol stalling problems. Parameters common to both the VT(S)-1 and the combined poll window/poll timer approaches are:
- RLC parameters:
  - o Last PDU in buffer polling
  - o Last PDU in retransmission polling
  - o Poll every n SDU where n=[4,6,8]
  - o Poll timer set to 250 ms
- System parameters:
  - o Iub round trip time of 200 ms
  - o Airlink frame error rate = 10%

---

[20] This work was performed by Dr. S. Abraham and Dr. M. Chuah while they were at Bell Laboratories

The mean object throughput is measured by averaging the object throughputs (not including the viewing time) within a data session while the maximum object throughput is measured by taking the maximum of the object throughputs within a data session. The mean page throughput is measured by averaging the page throughput (including the viewing time) within a data session while the maximum page throughput is measured by taking the maximum of the page throughputs within a data session. The maximum object throughput obtained is similar to the throughput obtained with FTP experiments discussed in Section 9.2.1. The results indicate that the mean page throughput with the "VT(S)-1" approach is similar to that achieved using the combined poll timer based approach. The combined poll timer based approach uses fewer uplink status messages.   However, it is noted that the VT(S)-1 approach is not part of the 3GPP standards, while the poll-timer/poll-window based approach uses features provided by the standards specifications.

Table 9-6. Results for the HTTP Experiments with RLC Deadlock Avoidance Mechanism

Expt	Num SDU	Avg Obj. Thpt	Max Obj. Thpt	Avg Page Thpt	Max Page Thpt	Mean Pkt Delay	Dnlink Status Bytes/page	Uplink Status Bytes/page
Poll Timer	4	18.4	40.3	14.5	32.5	1.3	0.47	0.61
	6	17.9	39.6	14.2	32.1	1.13	0.39	0.12
	8	18.6	40.7	14.2	31.4	1.37	0.39	0.14
VT(S)-1	4	18.4	39.8	14.6	32	1.23	0.49	0.48
	6	18.1	41.8	14.4	33.2	1.25	0.47	0.47
	8	18.3	39.7	14	29	1.15	0.42	0.15

9.3   Recommendations of TCP Configuration Parameters over 3G Wireless Networks

There are a few link layer characteristics of 3G networks that produce significant effects on TCP performance, namely [Ina03]:

1. Latency
    The latency of 3G links is high mostly as a result of the extensive processing required at the physical layer e.g. for FEC and interleaving, and as a result of the transmission delays in the radio access networks (which includes link-layer retransmission delays).

2. Varying data rates
    Both CDMA2000 and UMTS networks are based on CDMA technology. Data rate that is allocated to a user is dependent on the interference generated by other users from other cells, and mobility of the user. A user closer to the base station can be

allocated higher data rates than a user far away from the base station.

3. Delay spikes

A delay spike is a sudden increase in the latency of the communication path [Ina03]. 3G links are likely to experience delay spikes due to a couple of reasons, e.g., a long delay spike due to link layer recovery when a user drives into a tunnel or during an handover when the mobile terminal needs to exchange some signaling messages with the new base station before data can be transmitted in a new cell.

4. Intersystem handovers

In the initial phase of 3G deployment, 3G systems will be used as a "hotspot" technology in densely populated areas while the 2.5G system will provide lower speed data service in most areas. This creates an environment where a mobile can roam between 2.5G and 3G networks while keeping ongoing TCP connections. The intersystem handover will likely create a delay spike.

RFC3481 [Ina03] provides a set of recommendations for configuration parameters to support TCP connections over 3G wireless networks. We summarized the major recommendations below:

1. Appropriate Window Size

Appropriate window size based on the bandwidth delay product of the end-to-end path needs to be supported. The buffer size at the sender and the receiver should be increased to allow a large enough window.

2. Increased Initial Window

The traditional initial window value of one segment and the delayed ACK mechanism means that there will be unnecessary idle times in the initial phase of a data connection. Senders can avoid this by using a larger initial window size of up to four data segments (not to exceed 4 Kbytes) [All02]. The recommended suggestion is that the initital window (CWND) should be set to

$$CWND = min(4*MSS, max(2*MSS, 4380 \text{ bytes}))$$

3. Limited Transmit

RFC3042 [All01] explains the limited transmit feature which extends fast retransmit/fast recovery for TCP connections with small congestion windows that are not likely to generate the three duplicate acknowledgements required to trigger fast retransmit. If a sender has previously unsent data queued for transmission, the limited transmit mechanism allows the sender to send a new data segment in response to teach of the first two duplicate acknowledgments that arrive at the sender. This feature is recommended for TCP implementations over 2.5G/3G networks.

4. Path MTU Discovery
    TCP over 2.5G/3G implementations should implement Path MTU discovery. Path MTU discovery allows a sender to determine the maximum end-to-end transmission unit. This allows TCP senders to employ larger segment sizes instead of assuming the small default MTU.

5. Selective Acknowledgments (SACK)
    If an end-to-end path has a large bandwidth delay product and a high packet data rate, then the probability of having multiple segment losses in a single window of data increases. In such cases, SACK [Mat96] provides robustness beyond TCP-Tahoe and TCP-Reno.

6. Explicit Congestion Notification (ECN)
    TCP implementations over 2.5G/3G should support ECN. ECN [Ram01] allows a TCP receiver to inform the sender of congestion in the network by setting the ECN-Echo flag upon receiving an IP packet marked with the CE bit(s). The TCP sender can then reduce its congestion window. If intermediate routers along the end-to-end path do support the ECN feature, then there will be potential performance improvement.

Interested readers can refer to RFC3481 for more information and suggestions.

9.4   Some Proposed Techniques to Improve TCP/IP Performance in 3G Networks

Since wireless channels are subjected to higher error rates, link layer retransmission protocols, e.g., Radio Link Protocol (RLP) and Radio Link Control (RLC) are used in 3G1x [RLP], and UMTS [25.322] respectively to ensure that the packet loss probability on the wireless links will be less than 1% with local retransmissions. Although such techniques can mitigate losses, they often increase the delay variability seen by TCP. The way how packets are scheduled can affect the variability seen by TCP too. Intelligent channel state based scheduling [Bha96] that takes the quality of a user's perceived channel condition into considerations when the base station schedules data packets from various users, has been proposed. Another channel-state based scheduling approach is the proportional fair algorithm discussed in [Ben00] which can provide long-term fairness among different users. However, such channel state based scheduling schemes tend to increase the rate variability seen by TCP. The increase in delay and rate variability translates to bursty acknowledgement (ACK) arrivals at the TCP source. Such bursty ACK arrivals result in a burst of packets from the TCP source which could result in multiple packet losses when they arrive at a link with variable rate or delay. These multiple losses often degrade TCP throughput.

An ACK regulator is proposed in [Mun02] to improve the TCP performance in the presence of varying bandwidth and delay. The proposed technique can be implemented at RNC. The intuition behind the proposed algorithm is to avoid any buffer overflow losses until the congestion window has reached a predetermined threshold and beyond that allow only a single buffer overflow loss. This ensures that the TCP source operates mainly in the congestion avoidance phase exhibiting the classic saw-tooth behavior.

Simulation studies were performed [Mun02] for the proposed ACK regulator scheme. The results presented in [Mun02] indicate that with exponential forward delay and a fixed round trip delay, the throughput for TCP Reno (SACK) dropped by 30% (19%) respectively when the delay variance changes from 20 to 100. With the ACK regulator, the TCP throughput only drops by 8% when subjected to the same delay variance. Similarly, when subject to bandwidth variation, the ACK regulator scheme improves the throughput of TCP Reno by 15%. Interested readers can refer to [Mun02] for more detailed discussions of their simulation results.

Acknowledgments

The RLC/MAC work reported here was jointly carried out with the author's former Bell Laboratories colleagues Dr. S. Abraham, Dr. H. Su, and Dr. W. Lau. The call setup delay analysis and voice bearer delay analysis work was jointly done with S. Abraham, and E. Hernandez-Valencia from Bell Laboratories.

9.5   Review Exercises

(1) Assuming that the RRC connection request message is 200 bytes in size, a 64 Kbps data channel is used to deliver this message with a TTI of 20 ms, the RLC PDU packet error rate is 5%, and the maximum retransmission is set to 3, compute the probability of successfully transmitting the whole RRC connection request message and also compute the average time taken to deliver the whole message if it is delivered successfully.

(2) Analyze the simple analytical model provided in Section 9.2.1

(3) Extend the model developed in [Lak97] to analyze TCP/IP performance in 3G networks.

(4) Name the different techniques that have been proposed by researchers to improve the TCP performance in 2.5/3G networks.

9.6   References

[23.060] 3GPP TS23.060 draft v3.2.0 (2000-01). GPRS Service Description: Stage 2.

[25.322] 3GPP TS 25.322 "RLC Protocol Specifications," 1999.

[25.331] 3GPP TS 25.331, "RRC Protocol Specifications," 1999.

[3Gdelay] Vodafone Limited, "Access Delay Stratum Contribution," R3-00571.zip, September, 1999.

[All01] M. Allman et al,"Enhancing TCP's loss recovery using limited transmit," RFC3042, January 2001.

[All02] M. Allman et al, "Increasing TCP's Initial Window," RFC3390, October, 2002.

[Ben00] P. Bender et al, "A Bandwidth Efficient High Speed Wireless Data Service for Nomadic Users," IEEE Communication Magazine, July, 2000.

[Bha96] P. Bhagwat et al, "Enhancing Throughput over Wireless LANs using channel state dependent packet scheduling," Proceedings of IEEE Infocom, 1996.

[Dos98] B. Doshi et al, "Protocols, Performance and Controls for Voice over Wide Area Packet Networks," Bell Laboratories Technical Journal, Vol 3, No 4, Oct–Dec 1998.

[ETR102] ETR 102: August 1996 (GSM03.05 Version 4.1.0).

[G.114] ITU-T Recommendation G.114, One-way Transmission Time, 1998.

[GR253] Bellcore GR-253-CORE, Issue 2, December 1995.

[Her00] E. Hernandez-Valencia, M. Chuah, "Transport Delays for UMTS VoIP," Proceedings of WCNC, 2000.

[Ina03] H. Inamura et al, "TCP over 2.5G and 3G Wireless Networks," RFC3481, February, 2003.

[Lak97] T. Lakshman and U. Madhow, "The Performance of Networks with High Bandwidth-Delay Products and Random Loss," IEEE/ACM transactions in Networking, June, 1997.

[Luc99] Lucent Technologies, "Dedicated Voice Service Requirements for RT-EGPRS," ETSI SMG2 EDGE Workshop, Tdoc 2e99–439r1, 1999.

[Mat96] M. Mathis et al, "TCP Selective Acknowledgements Options," RFC2018, October, 1996.

[Mah97] B. Mah, "An Empirical Model of HTTP Network Traffic," Proceddings of Infocom 1997.

[Mun02] M. Chan, R. Ramjee, "TCP/IP Performance over 3G wireless links with rate and delay variation," Proceedings of Mobicom, 2002.

[NS] Network simulator http://www.isi.edu/nsnam/ns.

[Qin02] W. Lau, W. Luo, H. Su, O. Yue, Q. Zhang, "Analysis of UMTS Radio Link Control," Proceedings of Globecom, 2002.

[Ram01] K. Ramakrishnan, et al,"The addition of Explicit Congestion Notification (ECN) to IP," RFC3168, September, 2001.

[RLP] TIA/EIA/IS707-A-2.10, "Data Service Options for Spread Spectrum Systems: Radio Link Protocol Type 3," January, 2000.

[Su03] H. Su, Q. Zhang, "Methods for preventing protocol stalling in UMTS Radio Link Control," Proceedings of ICC, 2003.

Chapter 10

OVERVIEW OF WIRELESS LAN

10. INTRODUCTION

A wireless local area network (WLAN) is a communication system where a user connects to a LAN using the radio frequency (RF) technology. WLAN is designed as an alternative to the wired LAN to minimize the need for wired connections. It combines data connectivity with user mobility.

Wireless LAN functions similarly to a cellular system. As shown in Figure 10-1, each access point is a base station that transmits data between the wireless LAN and the wired network infrastructure. A single access point can support a small group of users and provide coverage of a radius of distance. Access points are connected to a wired network via an Ethernet hub or switch. End users access the wireless LAN through wireless LAN adapters, which are either built into laptops or added via PC cards. Users in the wireless LAN can seamlessly roam between access points without dropping their connections.

Wireless LANs have gained strong popularity [Lam96][Cro97][Zeh00] in a number of vertical markets, including health-care, academia, retail, hotels, and manufacturing. These industries have profited from the advantages offered by the wireless LANs. With wireless LANs, users can access the shared information anywhere in the LAN coverage area. Network administrators can set up or augment networks without installing new wires. The mobility support, installation flexibility, and simplicity provide service opportunities and cost benefits for wireless LANs over the wired networks.

The technologies for wireless LANs are specified by the IEEE 802.11 standards [802gro], which are often branded as "Wi-Fi" (Wireless Fidelity) supported by Wi-Fi Alliance [WiFi]. The Wi-Fi Alliance is a nonprofit international association funded in 1999. It is devoted to certify

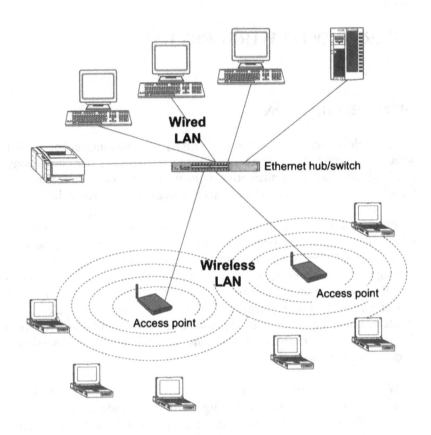

Figure 10-1 Wireless LAN Architecture

interoperability of wireless LAN products based on the IEEE 802.11 specifications. Currently the Wi-Fi Alliance has over 200 member companies from around the world. Since March of 2000, more than a thousand products have received the Wi-Fi certification and use the Wi-Fi logo.

Bluetooth is the technology for wireless Personal Area Network (PAN). It was developed by the Bluetooth Special Interest Group [Blue], which was founded in 1998 by Ericsson, IBM, Intel, Nokia and Toshiba. The name Bluetooth comes from King Harald Blatan (Bluetooth) of Denmark. Bluetooth is an open standard for short-range transmission of digital voice and data between mobile devices (laptops, personal digital assistants, phones, etc) and desktop devices. It provides up to 720 kbps data transfer within a range of 10 meters and up to 100 meters with a power boost.

Bluetooth transmits in the unlicensed 2.4 GHz band and uses a frequency hopping spread spectrum technique.

Infrared transmission (IrDA) is another wireless access technology for transmission of very short distance between two devices. IrDA stands for Infrared Data Association, a membership organization that was founded in 1993 [IrDA]. It is devoted to define specifications for infrared wireless communications. With IrDA ports, a mobile device such as laptop or PDA can exchange data with a desktop computer or use a printer without a cable connection. IrDA requires line-of-sight transmission similar to a television remote control. IrDA provides up to 115.2 kbps data transfer. High-speed extensions up to 4 Mbps for fast infrared data transfer is also defined.

10.1 Overview of 802.11 Wireless LAN

IEEE 802.11 standards include a family of specifications for wireless LANs developed by a working group of IEEE. The objective of the standard is "to develop a Medium Access Control (MAC) and Physical layer specification for wireless connectivity for fixed, portable and moving stations within a local area." There are currently five specifications in the family: 802.11, 802.11a, 802.11b, 802.11e, and 802.11g. In the following sections, we will describe the most widely implemented standards: 802.11a, 802.11b, and 802.11g.

10.1.1 Wireless LAN Architecture and Configurations

There are two basic configurations or operation modes for an 802.11 system, "*infrastructure*" mode and "*ad-hoc*" mode.

In "infrastructure" mode as shown in Figure 10-2, the wireless devices communicate to a wired LAN via access points. Each access point and its wireless devices are known as a Basic Service Set (BSS). An Extended Service Set (ESS) consists of two or more BSSs in the same subnet. Access points not only provide communication links between the wireless LANs and wired LANs but also manage wireless data traffic. Multiple access points can provide large coverage for an entire building or campus. The infrastructure mode allows wireless users to efficiently share network resources.

The "ad hoc" mode, also known as "peer-to-peer" mode as shown in Figure 10-3, is the simplest wireless LAN configuration. In "ad hoc" mode, wireless devices can communicate with each other directly and do not use an

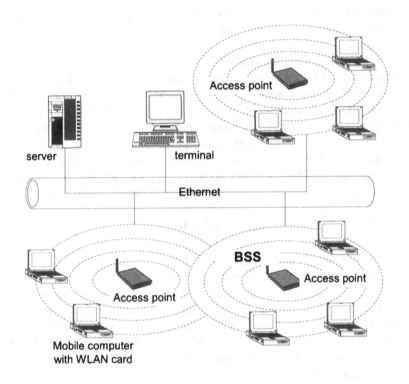

Figure 10-2 Wireless LAN Infrastructure Mode

access point. This is so called an Independent BSS (IBSS). These independent networks usually do not require any administration or pre-configuration. Access points can extend the range of "ad-hoc" WLANs via functioning as a repeater, effectively extending the distance between the devices.

10.1.2   802.11b

The 802.11b standard for wireless LAN, also known as Wi-Fi, is part of the IEEE 802.11 series of wireless LAN standards. 802.11b specification was ratified in 1999. It is the first commercial wireless LAN system and has been widely deployed in the world. It is backward compatible with 802.11 and can provide data rate from 1 to 11 Mbps in a shared wireless local area network.

802.11b operates in the 2.4-GHz Industrial, Scientific, and Medical (ISM) band. The ISM band is a part of the radio spectrum that can be used

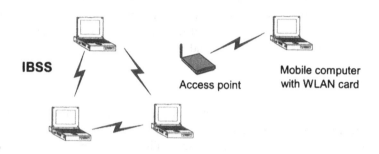

IBSS

Access point

Mobile computer
with WLAN card

Figure 10-3 Wireless LAN Ad-Hoc Mode

by anyone without a license in most countries. The first-generation WLAN standard referred to as 802.11b supports three physical layer specifications, namely Frequency Hopping Spread Spectrum (FSSS), Direct Sequence Spread Spectrum (DSSS), and Infrared (IR). The data rate was originally designed to be 1-2 Mbps and operates in the 2.4-GHz band. Since then several revisions have been added to increase the data rate. Currently, 802.11b can support up to 11 Mbps. The modulation scheme used in 802.11 has historically been phase-shift keying (PSK). The modulation method selected for 802.11b is known as complementary code keying (CCK), which allows higher data rates and is less susceptible to multi-path propagation interference. 802.11b offers data rate from 1 Mbps to 11 Mbps with a coverage distance of a hundred feet. The transmission speed depends on distance. The farther the distance between a transmitter and a receiver, the lower the bit rate in an 802.11 system. The data rates and modulation schemes in 802.11b are summarized in Table 10-1.

Table 10-1 Data Rate and Modulation Scheme in 802.11b

Data Rate (Mbps)	Modulation Scheme	Channel Coding
1	BPSK	Barker (11 chip)
2	QPSK	Barker (11 chip)
5.5	QPSK	CCK (8 chip)
11	QPSK	CCK (8 chip)

Like other 802.11 standards, 802.11b uses the carrier sense multiple access with collision avoidance (CSMA-CA) media access control (MAC) protocol for path sharing. A wireless station with a packet to transmit first listens on the wireless medium to determine if another station is currently transmitting (this is the carrier sense portion of CSMA-CA). If the medium is being used, the wireless station calculates a random backoff delay. Only after the random backoff delay elapses can the wireless station again listen for a transmitting station. By instituting a random backoff delay, multiple stations that are waiting to transmit do not end up trying to transmit at the same time (this is the collision avoidance portion of CSMA/CA).

10.1.3   802.11a

802.11a is a new specification for the next generation of enterprise-class wireless LANs. Compared to 802.11b, 802.11a has the advantages of better interference immunity and larger scalability. It can provide much higher speeds, up to 54 Mbps.

802.11a operates in the 5–GHz Unlicensed National Information Infrastructure (U-NII) band and occupies 300 MHz of bandwidth. The FCC has divided the total 300 MHz into three distinct 100 MHz domains, each with a different regulated maximum transmission power. The "low" band operates between 5.15 and 5.25 GHz, with maximum power output of 50 mW. The "middle" band operates between 5.25 and 5.35 GHz, with maximum power of 250 mW. The "high" band operates between 5.725 and 5.825 GHz, with maximum power of 1 W. The low and middles bands are more suited for in-building wireless transmission, while the high band is more suited for building-to-building transmission. The U-NII band is less populated than the ISM band and has less interfering signals from other wireless devices. Moreover, the 300 MHz of bandwidth is nearly four times the bandwidth available in the ISM band and thus can support higher data rates.

802.11a uses the Orthogonal Frequency Division Multiplexing (OFDM) as the transmission scheme. There are a total of 8 non-overlapping channels defined in the two lower bands, each of which is 20 MHz wide. Each of these channels is divided into 52 sub-carriers. Each sub-carrier is approximately 300 kHz wide. Information bits are encoded and modulated in each sub-carrier, and the 52 sub-carriers are multiplexed together and transmitted in "parallel". Therefore, high data rate is realized by combining many low rate sub-carriers. The transmission of multiple sub-carriers or sub-channels makes the network more scalable than other techniques. To combat the channel errors and improve the transmission quality, Forward Error Correction (FEC) was introduced in 802.11a. In addition to offering higher data rate and better scalability, another significant advantage of OFDM technique is that it improves the interference resistance over the multipath fading channels. Because of the low symbol rate of each sub-carrier, the effect of delay spread is reduced and thus the multipath interference is minimized.

The 802.11a standard requires the devices to support data rates of 6, 12, and 24 Mbps. Other data rates up to 54 Mbps are optional. These different data rates are the result of different modulation and coding scheme as shown in Table 10-2. The coverage provided by 802.11a system is similar to the coverage of 802.11b, but with a significantly higher data rate. The comparison of the data rates and coverage between 802.11b and 802.11a is shown in Figure 10-4.

802.11a uses the same MAC technique as 802.11b, i.e., CSMA-CA for access and collision avoidance. Although their MAC layer designs are the same, 802.11a and 802.11b have significant differences at the physical layer and operate at different frequency spectrum. Therefore, the two systems are not compatible, which makes the migration from 802.11b to 802.11a a bit more difficult. Some networks use both technologies concurrently.

802.11a has some similarities to HiperLAN standard. HiperLAN is a specification developed by ETSI for wireless transmission. It also uses the OFDM technology and operates in the 5 GHz frequency band. However, its MAC is different from 802.11a. HiperLAN uses TDMA technique instead of CSMA-CA. Since both systems operate in the N-UII band, 802.11a needs certification by ESTI so that it can coexist with HiperLAN.

Table 10-2 Modulation and Coding Scheme in 802.11a Standard

Data Rate (Mbps)	Modulation Scheme	Channel Coding Rate
6*	BPSK	½ convolutional code
9	BPSK	¾ convolutional code
12*	QPSK	½ convolutional code
18	QPSK	¾ convolutional code
24*	16 QAM	½ convolutional code
36	16 QAM	¾ convolutional code
48	64 QAM	2/3 convolutional code
54	64 QAM	¾ convolutional code
*: mandatory rates.		

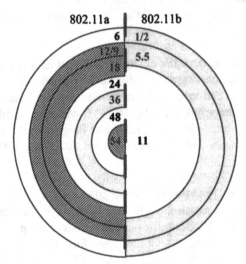

Values are data rate in Mbps. The coverage distance is approximate.

Figure 10-4 Data Rate Coverage Comparison Between 802.11b and 802.11a

802.11a is a very high-speed and highly scalable wireless LAN standard. It can offer data rate up to 54 Mbps to accommodate high-bandwidth applications. An 802.11a system has a similar coverage as 802.11b but provides a much higher throughput than an 802.11b system. It employs the OFDM technology and has eight non-overlapping channels, which is highly scalable and flexible for implementation. It operates in the 5 GHz spectrum that is less populated with less interference and congestion than the ISM band. The advantages of speed, scalability, and interference immunity have made 802.11a a potential solution for enterprise wireless LAN.

10.1.4   802.11g

802.11g is a recently approved wireless LAN standard by IEEE. It offers high data rate at up to 54 Mbps, which is comparable to the 802.11a standard. Most importantly, 802.11g provides backward compatibility to the widely implemented 802.11b standard. 802.11g has gained much interest among wireless users.

802.11g operates in the 2.4 GHz ISM band, the same as 802.11b. It uses OFDM technology that is also used by 802.11a. It supports high data rate of 6, 9, 12, 18, 24, 36, 48, and 54 Mbps, the same rate set as employed in

802.11a. For the purpose of backward compatibility with 802.11b, 802.11g also supports Barker code and CCK modulation offering data rate of 1, 2, 5.5 and 11 Mbps. Among the data rates, 1, 2, 5.5, 11, 6, 12, and 24 Mbps are mandatory for transmission and reception as specified by the standard. Similar to the 802.11b standard, 802.11g has only three non-overlapping channels and the new physical layer is called the Extended Rate Physical (ERP) layer.

802.11g uses the same MAC technique, CSMA-CA, as 802.11a and 802.11a. Each 802.11 packet consists of preamble, header, and payload. In the 802.11b standard, the long preamble (120 μs) is mandatory while the short preamble (96 μs) is optional. In the 802.11g standard, both short and long preambles are supported to improve the transmission efficiency and maintain backward compatibility.

The slot time is different for 802.11b and 802.11g systems. 802.11g uses a shorter slot time (9 μs) than the slot time (20 μs) used in 802.11b. Shorter slot time improves the throughput of the system. However for backward compatibility, 802.11g also supports the longer slot time.

The physical layer specification for 802.11g is almost the same as 802.11a. As a result, both systems should have similar capacity and performance. However in reality, the performance for 802.11g is different from 802.11a due to the following reasons.

1. 802.11g shares the same 2.4-GHz spectrum as 802.11b. When both 802.11b and 802.11g devices are present, proper coordination and management between the two different devices is necessary to avoid inference and collision. The performance impact on 802.11g devices can be significant if no protection is employed. On the other hand, 802.11a uses a different spectrum and thus does not need to coordinate with other wireless LAN users.
2. To maintain backward compatibility, 802.11g is required to use the 20 μs time slot as 802.11b. The use of 9 μs time slot is optional in 802.11g. In 802.11a, the time slot duration is 9 μs, which is more efficient than the 20 μs time slot.
3. The 2.4 GHz ISM spectrum used by 802.11g is more congested than the 5 GHz U-NII spectrum used by 802.11a. Thus 802.11g devices are subject to more interference sources, such as the microwave ovens, Bluetooth devices, cordless phones, and medical RF devices, etc. For 802.11a devices, since the 5 GHz spectrum does not overlap with an ISM band, there are fewer inference sources from external devices.

4. The number of available channels in the 2.4 GHz ISM band is less than that in the 5 GHz U-NII band. There are only three non-overlapping channels in 802.11g, compared with the 13 available channels in 802.11a. As a result, frequency reuse factor is smaller in 802.11g than 802.11a to meet the capacity and coverage requirement. Co-channel interference due to frequency reuse is higher in 802.11g than 802.11a.

5. The propagation path loss at 5 GHz band is higher than that at 2.4 GHz band, which favors 802.11g. The most important advantage of 802.11g over 802.11a is the backward compatibility with the legacy 802.11b devices.

10.2   802.11 Physical Layer

The original IEEE 802.11 standard recommends three different types of physical layers implementations:
- Frequency hopping spread spectrum (FHSS)
- Direct sequence spread spectrum (DSSS)
- Infrared (IR) light

In FHSS systems, the frequency at which data is transmitted changes in a pattern known to both transmitter and receiver. The data is sent on a given frequency for a fixed period of time, i.e., the *dwell time* in 802.11, and then switched to the next frequency for another fixed period of time. In the United States and European version of the 802.11 standard, maximum 79 frequencies are specified in the hopping set. The FH defined in 802.11 systems is slow FH since the change of frequency is slow compared to the transmission symbol rate. The standard specifies three different sets of hopping patterns, with 26 patterns in each set. Different hopping patterns allow multiple BSSs to coexist in the same geographical area to minimize the co-channel interference.

In DSSS systems, the original data symbol is modulated by a wideband spreading sequence. To an unintended receiver, the DSSS signal appears as low-power wideband noise. The desired receiver knows the spreading sequence and can recover the original data symbol from the wideband noise. For example, in the 802.11b standard, the original 1 Mbps data sequence is encoded using the binary phase shift keying (BPSK). Each encoded symbol is then spread by an 11-chip Baker sequence. The total bandwidth occupied by the system is thus 11 MHz.

In IR systems, the wavelength for transmission has a range from 850 to 950 nm. The IR band is designed for indoor environment. The modulation scheme for IR transmission is pulse position modulation (PPM).

The IEEE 802.11 working group allowed multiple physical layer designs to exploit the advantages of each of them. The disadvantage is that it brings the complexity of interoperability and increases the cost of developing multiple physical layers.

In the series of standards followed by the original 802.11 standard, a single physical layer was specified. In 802.11b, the physical layer uses the DSSS technology. For 1 Mbps and 2 Mbps data rate, Barker code is used for spreading and coding. For 5.5 Mbps and 11 Mbps data rate, CCK is used instead for better protection against the multipath interference. In 802.11a, the physical layer uses the OFDM technology. In 802.11g, the physical layer uses the OFDM but it also supports DSSS to be backward compatible with 802.11b. A hybrid CCK and OFDM method is an option in 802.11g. In this method, the header of a packet is transmitted in a single radio frequency (CCK) and the payload is transmitted in multiple frequencies (OFDM). This option is designed to avoid collisions with the 802.11b systems. The CCK header will alert the 802.11b devices that a transmission is happening.

All the 802.11 standards use the same MAC design, i.e., CSMA-CA. The details of the MAC layer are described in Chapter 11.

10.3 Capacity and Performance of 802.11 System

10.3.1 Coverage and Throughput Performance

The traditional path loss and propagation channel model for cellular systems can be applied for wireless LAN systems. Since most of the wireless LAN is in indoor environment, a two-slop path loss model with a breakpoint can be used. Path loss is modeled as two straight lines with different slops intersecting at the break point. When the distance is smaller than the break point, the free space path loss is represented by

$$L_{path} = 20\log_{10}\left(\frac{4\pi d}{\lambda}\right)$$

(10-1)

where L_{path} is the free space path loss in dB; d is the distance, and λ is the wavelength. When the distance is larger than the break point, the path loss depends on the propagation environment and can be represented by

$$L_{path} = L_{path\_b} + 10 \cdot \alpha \cdot \log_{10}\left(\frac{d}{d_b}\right)$$

$$(10\text{-}2)$$

where L_{path_b} is the path loss at the break point d_b; α is the path loss exponent. The break point depends on the antenna height, the first Fresnel zone clearance, and the transmitted frequency. α depends on the propagation environment. Typical values of d_b and α can be chosen as: $d_b = 10$ meters, and $\alpha = 3.2$, which represent a medium obstructed office environment.

In addition to the free space path loss, multipath and shadow fading will have significant effects on a wireless LAN system. The multipath effect can be modeled as a Rayleigh fading channel with a 50-ns delay spread. The shadow fading can be modeled as a lognormal distributed random variable with an 8 dB standard deviation.

802.11b and 802.11g devices operate in the same frequency band. The path loss and fading model would be similar for the two systems. However, since 802.11a operates in a different frequency band, the effect of path loss will be different from 802.11b/g systems. Taking the free space path loss model, the difference in path loss between the 802.11a device and 802.11b/g device is approximately

$$\Delta L_{path} = 20\log_{10}\left(\frac{4\pi d}{\lambda_1}\right) - 20\log_{10}\left(\frac{4\pi d}{\lambda_2}\right)$$

$$= 20\log_{10}\left(\frac{f_1}{f_2}\right) = 20\log L_{10}\left(\frac{5}{2.4}\right) = 6.4$$

$$(10\text{-}3)$$

where f_1 and f_2 are the frequency of the U-NII, and ISM band respectively. Therefore, the path loss favors the 802.11b/g devices. In other words, the received signal strength at the 802.11b/g devices will have less loss (6.4 dB) than the 802.11a devices.

The data rate of 802.11 systems is distance dependent. The transmission data rate is chosen based on the achievable signal strength at the receiver to meet a certain link quality. A rate fallback algorithm is designed to select the transmission rate based on the distance and propagation environment. The selection criterion is usually based on packet error rate. 10% packet error rate is often used as the fallback threshold. The current data rate will be selected until the packet error rate reaches 10%. At this point, the system will reduce the data rate to the next lower data rate in the available rate set.

Both simulations and measurements have been conducted to evaluate the coverage and data rate performance for 802.11 systems [Che01][Pro01]. Table 10-3 summarizes the data rate versus distance for different wireless LAN systems. The main observations that can be drawn are:

- 802.11a can provide coverage comparable to 802.11b up to 225 feet in a typical office environment.
- The data rate of 802.11a is approximately 2 to 5 times better than 802.11b within the coverage distance.
- 802.11g can provide coverage comparable to 802.11b as well.
- The data rate of 802.11g in a single cell is higher than 802.11a because of the better frequency dependant path loss.

At the maximum distance, both 802.11a and 802.11g offer a 6 Mbps data rate compared to the 2 Mbps for 802.11b. At closer distance, the rate improvement for 802.11a and 802.11g is much more evident, i.e., 54 Mbps versus 11 Mbps for 802.11b.

Table 10-3 Data Rate versus Distance for Different 802.11 Systems

802.11a		802.11b		802.11g	
Data rate (Mbps)	Distance (feet)	Data rate (Mbps)	Distance (feet)	Data rate (Mbps)	Distance (feet)
54	24	11	115	54	42
48	35	5.5	180	48	47
36	83	2	225	36	65
24	88			24	85
18	133			18	107
12	170			12	135
6	225			6	180

Figure 10-5 A Simplified Frame Structure of an IEEE 802.11 Data Packet.

The value of data rate provides us insights on how the wireless LAN systems trade rate for coverage. Another important performance metric is the actual user perceived throughput. Throughput is defined as the ratio of the transmitted information bit to the transmission time. It is the actual rate of the information bits that can be transmitted by taking account of various overheads. Figure 10-5 shows a simplified frame format of an 802.11 data packet. A complete data packet consists of three segments. The first segment is called the preamble, which is used for signal detection, timing/frequency acquisition and synchronization. The second segment is called the overhead, which contains information of data rate, packet length, and addresses. The third segment is called the MAC protocol data unit (MPDU), which contains the actual information bits. The combination of preamble and overhead is called the physical layer convergence protocol (PLCP) field in an 802.11 packet.

Throughput depends on a number of factors: the transmission data rate, protocol overhead, MAC efficiency, preamble, and packet size. Other factors such as the contentions among multiple users, retransmission due to collisions, and higher layer protocol efficiencies (TCP/IP) can also affect the achievable throughput. As a result, the actual user perceived throughput is much less than the transmission data rate. For example, in an 802.11b data packet, the preamble has 144 bits, and the overhead has 48 bits. The preamble and overhead are transmitted at 1 Mbps and thus results in a 192 μs overhead. The actual data transmission rate is one of the defined possible data rates. The overall overhead is in the vicinity of 50%. The throughput is about 5 Mbps for an 11 Mbps data rate in 802.11b, and 30 Mbps for a 54 Mbps data rate in 802.11a and 802.11g. The detail discussion on MAC layer design and performance will be provided in Chapter 11.

10.3.2   Impact of Co-Channel Interference on System Capacity

The discussion in the previous section emphasizes on the link budget, where the performance between two nodes (AP to a mobile or mobile to mobile) is described. System capacity is an important metric that is most widely used for performance evaluation. System capacity usually refers to the aggregate throughput of an entire wireless LAN system comprised of multiple cells and a number of users.

In a single cell deployment, if there is only one mobile station, the system capacity is equivalent to the throughput received by the mobile station. For multiple mobile stations in a cell, the system capacity is the average cell throughput among all the mobile stations. A mobile station that is closer to the AP will have a higher throughput, while a mobile station that is in the edge of the cell will have a lower throughput. The average cell throughput is the aggregate throughput divided by the number of mobiles, assuming equal sharing among all the mobile stations. Based on the measured results discussed in previous section, the average cell throughput of an 802.11b system with a 225 feet cell radius is approximately 3.1 Mbps. The average cell throughput of an 802.11a system is approximately 9.4 Mbps, which is 3 times better than the 802.11b system. For a cell radius of 65 feet, the average cell throughput for 802.11a is almost 5 times better than 802.11b, i.e., 23 Mbps versus 5 Mbps.

In a multiple cell deployment, interference from other cells will have significant impact on the system capacity. In 802.11b systems, since there are only three channels available, the channel frequency has to be reused every three cells. Cells sharing the same channel frequency will cause interference to each other and reduce the throughput. In 802.11a systems, there are eight channels available. The impact of co-channel interference is less severe than the 802.11b systems. Figure 10-6 illustrates a three-tier (57 cells) cell layout and channel allocation for 802.11a and 802.11b systems. For 802.11b system, the frequency reuse factor is 3. Each cell has at least one adjacent cell that shares the same frequency and the average number of co-channel interfering cells is 1.67. For 802.11a system, the frequency reuse factor is 8. There is no interfering cell in the first and second tier. Interfering cell only exist in the third tier.

System level simulation can be used to evaluate the cell throughput under the co-channel interference. In the early stage of the IEEE wireless LAN standardization group, a system capacity model was proposed to evaluate the throughput. Two mechanisms were proposed to model the effect of co-channel interference on system capacity. The Clear Channel Assessment (CCA) method [Ish98] models the reduced throughput when the access point inside a particular cell has to wait until the channel is available for transmission. The other method, "Hidden Cell" models the decrease in

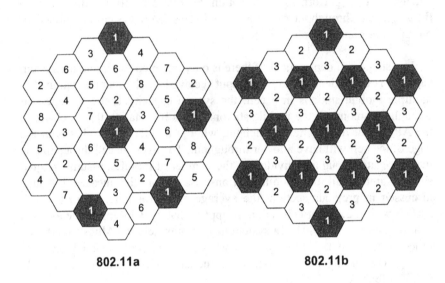

802.11a   **802.11b**

Figure 10-6 Co-Channel Interfering Cells in 802.11a and 802.11b Systems

throughput when a transmission from undetected cells corrupts other transmissions.

To quantify the co-channel interference, a parameter called the co-channel interference (CCI) immunity is defined, which is given as the carrier to interference ratio as

$$CCI = \frac{P_{carrier}}{P_{inter}} \qquad (10\text{-}4)$$

To guarantee a certain level of quality of service such as the packet error rate (e.g., PER below 10%), it is necessary to require that the CCI immunity be satisfied across the entire cell. The CCA scheme provides a method to detect any intolerable level of CCI. In the CCA scheme, a CCA threshold is defined, which is the minimum intolerable interference level derived from the CCI immunity,

$$P_{cca\_i} = P_{cell\_edge} - CCI_{immunity} \ (\text{dB}) \qquad (10\text{-}5)$$

where P_{cell_edge} is the received carrier power level at the edge of the cell. It is required that a transmitter should be able to detect the minimum intolerable interference threshold P_{t_CCA} by the CCA scheme. The threshold defined in (10-5) is an ideal CCA threshold. In practical system, the carrier power level should be above the minimal carrier sense level. Thus the practical CCA threshold can be set as

$$P_{cca\_t} = \max\left(P_{cell\_edge}, P_{min\_carrier\_sense}\right) - CCI_{immunity} \; (dB) \qquad (10\text{-}6)$$

An example of the CCA threshold and associated parameters are shown in Table 10-4.

Previous studies have shown that in a multiple cell deployment, the 802.11a systems can achieve much higher throughput (almost eight times higher) than the 802.11b systems. This throughput improvement results from the higher data rate 802.11a can offer, and also the fact that there is less co-channel interference because of increased number of available channels in 802.11a.

Table 10-4 Parameters in the CCA Threshold

Parameters	Value
Minimum sensitivity	−77 dBm
CCI immunity	9 dB
Minimal carrier sense level	−82 dBm
Ideal CCA threshold	−86 dBm
Practical CCA threshold	−82 dBm
Limited P_{cell_edge}	−73 dBm

For the 802.11g systems, since there are only three available channels as in 802.11b, there is not much gain in throughput due to the effect of co-channel interference. As a result, the throughput of 802.11g is lower than that of 802.11a in a multicell environment.

10.3.3 Performance of Mixed 802.11g and 802.11b Systems

Although the benefits of high speed and backward compatibility with the legacy 802.11b devices have made 802.11g the most promising next generation wireless LAN technology, there are some caveats, which cause

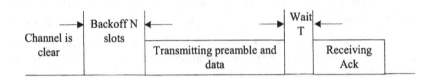

Figure 10-7 802.11 Packet Channel Access and Time Slots

performance degradation, especially in a mixed 802.11b and 802.11g environments.

For an 802.11 device to send a data packet, it needs to access the channel according to the MAC protocol. The 802.11 MAC layer uses the CSMA-CA, which attempts to avoid having different devices transmitting at the same time. A device can transmit data only when no other devices are transmitting to avoid any collision. As shown in Figure 10-7, when a device detects that the channel is clear, it waits for a random period time before sending the data. This will reduce the probability that another device will attempt to transmit at the same time. This random period of time is called the *backoff* time. After sending the data packet, the device needs to wait for acknowledgment before it can start its next transmission.

During the channel access procedure, the time is counted in a basic unit that is called the "slot time". The slot time value depends on the physical layer framing format, transmission rate and transmitter/receiver structure, etc. The slot time in 802.11b is defined to be 20 μs. In 802.11a, the slot time is defined to be 9 μs. In 802.11g, both slot times are supported. In the MAC layer, the access contention and random backoff time unit is based on slot. Shorter slot time has less transmission overhead and thus improves the efficiency. In a mixed 802.11b and 802.11g scenario, however, 802.11g devices have to use the longer slot time (20 μs) to fairly compete with the 802.11b devices. This is needed to avoid favorable treatment for devices with shorter time slots transmitting first all the time. If different slot times were used, the 802.11g devices with the smaller slot time would have preferable access to the wireless medium. Therefore 802.11g devices operating in the same network with 802.11b devices have to use the 20 μs slot time. The use of longer slot time diminishes the throughput gain in 802.11g. Compared with an 802.11a system or a 802.11g only system, the reduction in throughput using the longer slot time can be as much as 20% depending on the data rate the device is transmitting at.

OFDM signals transmitted by the 802.11g devices will not be detected by the 802.11b devices. 802.11b devices will incorrectly assume that the medium is free for transmission. This leads to collision and degrades the throughput. To prevent this from happening and improve the performance, the 802.11g standard requires the devices to employ protection mechanisms in a mixed 802.11b/g environment. Two protection mechanisms are specified in the standard: Request-To-Send (RTS)/Clear-To-Send (CTS) and CTS-to-self.

In the RTS/CTS scheme, the device intended for transmission sends a RTS message to the destination node. The destination node then sends a CTS message back to the sending device indicating that the RTS message was received and the sender could send the data packet. This CTS message is broadcasted to all the devices on the network so that other devices are aware that a transmission is occurring and would cease or delay their own transmission for a period of time, which is defined by the CTS message. After receiving the CTS message, the sender will send data and wait for acknowledgment if the transmission is successful. This RTS/CTS handshaking mechanism will result in fewer collisions.

In the CTS-to-self scheme, the access point will send a CTS message when it needs to send a downlink packet without receiving a RTS message first.

Both RTS/CTS and CTS-to-self mechanisms are used to reduce the collisions in packet transmission. However, the cost of these protections is that it takes longer time to transmit the same amount of data, and thus reduces the effective throughput. The throughput reduction in this case can be as much as 30%.

The 802.11g standard has drawn great interest in the wireless LAN industry. However, the deployment of this new technology is not as straightforward as it may first appear. Careful system design and coordination of the 802.11g devices with the legacy 802.11b devices are necessary to achieve the best system performance.

10.4   802.16 and Future Wireless LAN Technology

IEEE 802.16 is a standard for broadband wireless access [802.16]. The first IEEE 802.16 standard, published in April 2002, defines the Wireless MAN Air Interface for wireless metropolitan area network (MAN). It is designed as a high-speed, low-cost, and a scalable solution to extend the fiber optic backbones. These systems are meant to provide network access in the last mile to homes, small businesses, and commercial buildings as an alternative to traditional wired connections.

802.16 supports point-to-multipoint architecture in the 10 to 66 GHz range, transmitting at data rates up to 120Mbps. At those frequencies, transmission requires line-of-site, and roofs of buildings provide the best mounting locations for base and subscriber stations. The base station connects to a wired backbone and can transmit over the air up to 30 miles to a large number of stationary subscriber stations.

IEEE 802.16a, published in January 2003, specifies the 2 to 11 GHz operation frequency to be compatible with the 802.11 standards. It can accommodate non line-of-site access over these licensed and unlicensed low frequencies. The physical layer in 802.16a uses OFDM, which is similar to 802.11a and 802.11g standards.

The MAC layer in 802.16 is very different from the 802.11 MAC. The 802.16 MAC uses TDMA technique and can dynamically allocate uplink and downlink bandwidth to mobile stations. It supports many different physical layer specifications in both licensed and unlicensed bands.

The next step for the 802.16 working group is to add portability and mobility to the standard. In March 2002, the working group began the 802.16e Study Group on Mobile Broadband Wireless Access. This group will address many different mobility issues, including providing connectivity to moving vehicles within a base station's sector.

IEEE 802.20 working group [802.20] was established in December 2002 to develop specifications for the physical layer and medium access control layer that is optimized for the transport of IP based services. *"The goal is to enable worldwide deployment of affordable, ubiquitous, always-on and interoperable multi-vendor mobile broadband wireless access networks that meet the needs of business and residential end user markets."*[802.20]

IEEE 802.20 will operate in the licensed bands below 3.5 GHz. It requires the support of peak rate per user in excess of 1 Mbps. It also supports mobility with various mobile speeds up to 250 Km/h in a MAN environment. The target for 802.20 is to achieve better spectral efficiencies, higher sustained user data rates and larger numbers of active users than what can be achieved by the existing mobile systems.

The wireless access technology is evolving and future standard 802.11n is targeted to provide up to 100 Mbps broadband access. The availability is not expected until 2005 – 2006.

10.5   Review Exercises

1. Build a Monte-Carlo simulation to model the path loss and propagation model in a single user environment for 802.11b and 802.11a systems, using the parameters specified in section 10.3.1.
2. Calculate the overall overhead for an 802.11b packet, assuming a 1500- byte packet payload.
3. Calculate the added overhead when using the RTS/CTS protection mechanism and the resulted throughput reduction.
4. Use the model specified in [Ish98] to evaluate the co-channel interference effect on system throughput.

10.6   References

[802gro] IEEE 802.11 working group, http://grouper.ieee.org/groups/802/11
[802.11] IEEE standard P802.11, Wireless Medium Access Control and Physical Layer Specification.
[802.11a] IEEE standard P802.11a, Wireless Medium Access Control and Physical Layer Specification.
[802.11b] IEEE standard P802.11b, Wireless Medium Access Control and Physical Layer Specification.
[802.11g] IEEE standard P802.11g, Wireless Medium Access Control and Physical Layer Specification.
[802.16] IEEE 802.16 Standard Specification.
[802.20] IEEE 802.20 Standard Specification.
[Blue] Bluetooth Special Interest Group, http://www.bluetooth.com
[Che01] J. Chen, J. Gilbert, "Measured Performance of 5-GHz 802.11a Wireless LAN Systems," Atheros Communications, Inc.
[Cro97] B. Crow, I. Widjaja, J. Kim, P. Sakai, "IEEE 802.11 Wireless Local Area Networks," IEEE Communication Magazine, September 1997.
[IrDA] Infrared Data Association, http://www.irda.org
[Ish98] K. Ishii, "General Discussion of Throughput Capacity," IEEE 802.11-98, April 23, 1998.

[Lam96] R. LaMaire, A. Krishna, P. Bhagwat, J. Panian, "Wireless LANs and Mobile Networking: Standards and Future Directions," IEEE Communication Magazine, August 1996.

[Pro01] "A Detailed Examination of the Environmental and Protocol Parameters that Affect 802.11g Network Performance," White paper, Proxim Corporation.

[Zeh00] A. Zehedi, K. Pahlavan, "Capacity of a Wireless LAN with Voice and Data Services," IEEE Trans. on Communi., pp 1160-1170, Vol. 48, No. 7, July 2000.

[WiFi] Wi-Fi Alliance, http://www.wi-fi.org

[WiMAX] WiMAX Forum, http://www.wimaxforum.org

Chapter 11

MAC AND QOS IN 802.11 NETWORKS

11. INTRODUCTION

In recent years, much interest has been directed to the design of wireless networks for local area communications [Des96]: IEEE 802.11 study group was formed to recommend an international wireless local area network standard. The final version of the standard [802.11] provides detailed medium access control (MAC) and physical layer (PHY) specification for WLANs. 802.11b wireless LANs have grown rapidly in popularity these past 3 years. Wireless LANs provide high sustained data rates economically because they operate in unlicensed spectrum. Wireless LAN islands of coverage often referred to as public hotspots are being deployed in hotels, university campuses, airports, and cafes to large numbers of computer users including enterprise users. A new high-speed physical layer, the IEEE 802.11a PHY [802.11a] based on Orthogonal Frequency Domain Multiplexing (OFDM) technology has been developed to extend the existing IEEE 802.11 standard in the 5 GHz U-NII bands. As discussed in Chapter 10, 802.11a standard can support 8 different PHY modes with data rates ranging from 6 Mbps to 54 Mbps. 802.11a standard is not backward compatible to 802.11b. Thus, another standard 802.11g [802.11g] has been proposed that is backward compatible to 802.11b radios. 802.11g supports a data rate that ranges from 11 to 22 Mbps. There are two different medium access protocols specified in the standard [802.11]: the basic access mechanism, called the Distributed Coordination Function (DCF), based on the Carrier-Sense Multiple Access with Collision Avoidance (CSMA/CA) and a centrally controlled access mechanism, called the Point Coordination Function (PCF).

In Section 11.2, we describe the basic access method and Request-To-Send(RTS)/Clear-To-Send(CTS) mechanisms of the Distributed Coordination Function (DCF). We discuss Point Coordination Function in Section 11.3. Then, in Section 11.4, we discuss the performance evaluation of 802.11b DCF. We also discuss the performance anomaly that has been observed in 802.11b DCF operations. As in the cellular network, one is interested in understanding how TCP applications perform in wireless LAN environments and whether all users running TCP applications will be fairly treated in wireless LANs. We discuss such issues in Section 11.4. In Section 11.5, we discuss how voice services can be supported in 802.11b networks running PCF and summarize an analytical model proposed by some researchers for evaluating the voice capacity in WLANs. In Section 11.6, we discuss 802.11e – the new 802.11 standard that incorporates QoS features into WLANs. We discuss both enhanced DCF (EDCF) and hybrid coordination function (HCF) and presented some simulation results that some researchers have done in this area. We conclude this chapter by discussing two related MACs: (1) 802.11-based Cellular MAC and (2) 802.15.3 – a new MAC standard for high data rate wireless personal area networks.

11.1   802.11 Distributed Coordination Function

The IEEE 802.11b standard [802.11] defines two access methods: the distributed coordination function (DCF) that uses Carrier Sense Multiple Access (CSMA)/Collision Avoidance (CA) to allow users to contend for the wireless medium, and the Point Coordination Function (PCF) which provides uncontended access via arbitration by a Point Coordinator which resides in the Access Point (AP). The DCF method provides best effort service whereas the PCF guarantees time-bounded service. Both methods may coexist, a contention period following a contention-free period. PCF would be well-suited for real-time traffic but most of the currently available 802.11 products do not implement the PCF method.

The DCF access method as shown in Figure 11-1 is based on the CSMA/CA principle. A host wishing to transmit first senses the channel and waits for a period of time referred to as DIFS (Distributed Interframe Spacing) and then transmits if the medium is still idle. A station using the DCF has to follow two access rules: (1) the station shall be allowed to transmit only if its carrier sense mechanism determines that the medium has been idle for at least DIFS time; and (2) in order to reduce the collision probability between multiple stations, the station shall select a random backoff interval after deferral or prior to attempting to transmit again immediately after a successful transmission. If the packet sent is correctly received, the receiving host sends an ACK after another fixed period of time referred to as SIFS (Short Interframe Spacing). SIFS is designed to be smaller than DIFS so as to give a higher priority to the ACK frame. If this ACK is not received by the sending host, a collision is assumed to have

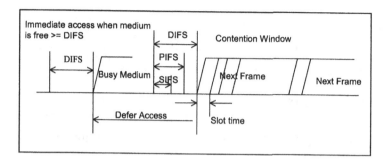

Figure 11-1. Basic Access Method ©IEEE Copyright 1997

occurred. The sending host attempts to resend the packet when the channel is free for a DIFS period augmented by a random backoff time.

The time immediately following an idle DFS is slotted, and a station is allowed to transmit only at the beginning of each slot time. The slot time size, σ, is set equal to the time needed at any station to detect the transmission of a packet from any other station. The slot time value depends on the physical layer (refers to Table 11-1), and accounts for the propagation delay, the switching time from the receiving to the transmitting state, etc.

DCF also adopts an exponential backoff scheme. At each packet transmission, the backoff time is uniformly chosen in the range $(0, w-1)$. The value w is called the contention window. At the 1st transmission attempt, w is set equal to CW_{min} (the mimimum contention window). After each unsuccessful transmission, w is doubled, up to a maximum value of CW_{max}. The values CW_{min} and CW_{max} specified in the standard [802.11] are physical layer specific and are summarized in Table 11-1.

A large network packet may be partitioned into smaller MAC frames before being transmitted. This process is called fragmentation. Fragmentation creates MAC protocol data units (MPDUs) that are smaller than the original MAC Service Data Unit (MSDU) length to improve the chance of successful fragment transmissions. Each fragment is sent as an independent transmission and is acknowledged separately. Once a station has contended for the medium, it shall continue to send fragments with a SIFS gap between the ACK reception and the start of subsequent fragment transmission until either all the fragments of a single MSDU have been sent or an ACK frame is not received. If any fragment transmission fails, the station shall attempt to retransmit the failed fragment after the backoff

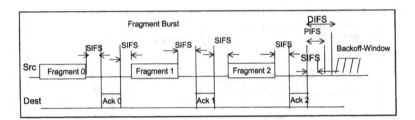

Figure 11-2. Timing Diagram of Successful Fragment Transmission ©IEEE CopyRight 1997

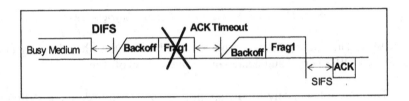

Figure 11-3. Timing Diagram of Fragment Transmission Failure and Retransmission ©IEEE CopyRight 1997

Table 11-1. Slot Time, CWmin, CWmax for 3 PHY Specifications

PHY	Slot Time	Cwmin	Cwmax
FHSS	50 us	16	1024
DSSS	20 us	32	1024
IR	8 us	64	1024

procedure. The timing diagrams for fragment transmissions are shown in Figures 11-2 and 11-3.

Apart from the basic 2-way handshake access mechanism, the 802.11 DCF also specifies a 4-way handshake access mechanism called RTS/CTS. A station that wants to transmit a packet, waits until the channel is sensed idle for a DIFS, follows the backoff rules explained above, and then transmits a short frame called Request to Send (RTS). When the receiving station detects the RTS frame, it responds after a SIFS, with a clear to send (CTS) frame. The transmitting station is allowed to transmit its data packet only if the CTS frame is correctly received. The RTS/CTS frames carry the information of the length of the packet to be transmitted. This information is read by any listening station, which then updates a network allocation vector (NAV) containing the information of the period of time in which the channel

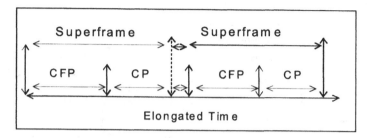

Figure 11-4. Timing Diagram for Superframe Consisting of CFP/CP

will remain busy. Therefore, when a station is hidden from either the transmitting or the receiving station, by detecting either the CTS or RTS frame, the node can avoid sending packets that will cause collisions.

11.2   802.11 Point Coordination Function

The centrally coordinated, access mechanism of the IEEE 802.11 MAC called the PCF, uses a poll-and-response protocol to control the access to the shared wireless medium. It makes use of the priority interframe space (PIFS) to seize and maintain control of the medium. The period during which the PCF is operated is called the contention-free period (CFP). The CFP alternates with the contention period. The sum of the two periods is called a superframe and is as shown in Figure 11-4. As shown in Figure 11-4, it may happen that a station begins to transmit a frame just before the end of the CP, hence elongating the current superframe and shortening the next CFP.

The Access Point performs the function of the point coordinator (PC). The PC maintains a list of stations that need to be polled. The PC tries to gain control of the medium at the beginning of the CFP after sensing the medium to be idle for a PIFS period. Once the Point Coordinator (PC) has control of the medium, it starts transmitting downlink traffic to stations. Alternatively, the PC can also send contention-free poll (CF-Poll) frames to those stations that have requested contention-free services for their uplink traffic. On receiving a poll, the station transmits its data after a SIFS interval if it has uplink traffic to send. Otherwise, it will respond with a NULL frame, which is a data frame without any payload. To utilize the medium more efficiently during the CFP, both the acknowledgment (CF-Ack) and the CF-Poll frames can be piggybacked onto data frames.

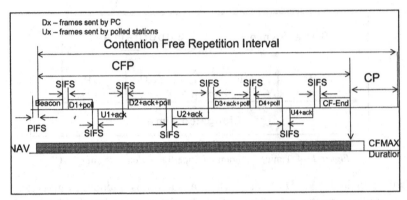

Figure 11-5. Timing Diagram of Successful Transmissions Under PCF Operation
©IEEE Copyright 1997

Figure 11-5 shows the detailed PCF operation where the Point Coordinator (PC) broadcasts a beacon every superframe (also referred to as contention-free period repetition interval) and transmits a downlink frame and poll for any uplink traffic to each station on its polling list. If there is uplink traffic, the station will send the uplink traffic and piggyback an ACK for the received downlink frame. A CF-End message will be sent to inform the stations of the ending of the contention free period so that they can contend during the contention period.

Consider an example of uplink data frame transmission [Qia02]. The PC first sends a CF-Poll to the wireless station and waits for an uplink data frame. As shown in Figure 11-6(a), if a data frame is received correctly within SIFS time, the PC will send a CF-Ack+CF-Poll frame that allows the next uplink data frame transmission. If a data frame is received in error, determined by an incorrect Frame Check Sequence (FCS), the PC will send a CF-Poll asking for the retransmission as shown in Figure 11-6(b). However, if no reply is received within a SIFS interval possibly due to an erroneous reception of the preceding CF-Poll frame by the polled station, the PC will reclaim the medium and send its next CF-Poll after a PIFS interval from the end of the previous CF-Poll frame as shown in Figure 11-7. In this case, the PC will not be confused with the scenario where the polled station has nothing to transmit because a NULL frame is expected under that circumstance. The PC may choose to re-poll the same station instead of skipping to poll the next station on its polling list.

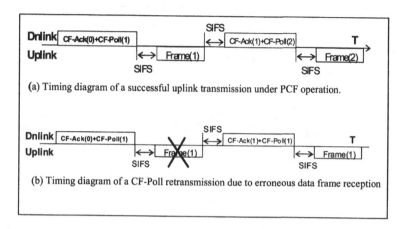

Figure 11-6. Timing Diagram of a Successful and Unsuccessful Uplink Transmission Under PCF ©IEEE Copyright 2002 [Qia02]

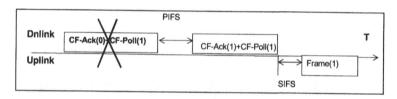

Figure 11-7. Timing Diagram of a CF-Poll Retransmission ©IEEE Copyright 2002 [Qia02]

11.3 Performance Evaluation of 802.11 DCF for Data Users

Wireless LAN network administrators are always interested in knowing the traffic engineering guidelines on how many data users can use the service of the same access point (using 802.11 DCF). Thus, it is of interest to study the performance of 802.11 DCF for data users. Several papers on such performance studies have been published, e.g. [Cal98][Bia00]. Here, we summarize the work done in [Heu03] since we would like to explain the performance anomaly that has been observed by the authors in [Heu03] in 802.11b-based networks.

11.3.1 Performance Evaluation of 802.11b DCF

Here, we summarize the analytical approach provided in [Heu03]. Let us consider a single host transmitting a data frame in a 802.11b cell. If we

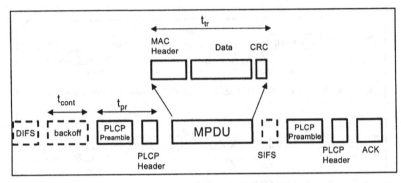

Figure 11-8. Successful Transmission of a Frame in 802.11b Network [Heu03] ©IEEE
Copyright 2003

ignore the propagation time, the time it takes to transmit a data frame successfully (assuming no collision) is given by

$$T = t_{tr} + t_{ov} \qquad (11\text{-}1)$$

where the constant overhead is given by

$$t_{ov} = DIFS + t_{pr} + SIFS + t_{pr} + t_{ack} \qquad (11\text{-}2)$$

where t_{pr} is the PLCP (Physical Layer Convergence Protocol) preamble and header transmission time, SIFS = 10 μs (using DSSS PHY layer), t_{ack} is the MAC acknowledgment transmission time (which is 10 μs if the selected rate is 11 Mbps since the ACK length is 112 bits), and DIFS = 50 μs. t_{tr} is the frame transmission time and it varies with the bit rate used by the host. In addition, t_{pr} is also a function of the bit rate used by the host, e.g. t_{pr} =192 μs if the user uses 1 Mbps data rate since the long PLCP header is used but if the data rate used is 2, 5.5 or 11 Mbps, then t_{pr} = 96 μs (short PLCP header). For bit rates greater than 1 Mbps, the proportion p of the useful throughput measured above the MAC layer with a frame size of 1500 bytes of data (with 34 bytes overhead) will be

$$p = \frac{t_{tr}}{T} * \frac{1500}{1534} = 0.7$$

So, a single host sending long frames over an 11-Mbps radio channel can at most have a throughput of 7.74 Mbps. When multiple hosts attempt to transmit, the channel may be sensed busy and hosts enter into a collision avoidance phase, i.e., every host executes the exponential backoff algorithm. Each host waits for a random interval distributed uniformly between [0,W]x

SLOT. The congestion window W varies between $CW_{min} = 31$ and $CW_{max} = 1023$, the value of SLOT is 20 µs for 802.11b with DSSS as PHY layer. The host that picks the smallest interval starts transmitting and the others stop the counting down until the transmission is over. When a host is involved in collisions, it doubles W up to CW_{max}.

When there are N-1 other hosts, the overall transmission time is given by [Heu03]

$$T(N) = t_{tr} + t_{ov} + t_{cont}(N) \tag{11-3}$$

The analytical formula for $t_{cont}(N)$ is difficult to derive (see [Cal98],[Bia00]). A simple approximation suggested in [Heu03] is given below:

$$t_{cont}(N) \approx SLOT * \frac{(1 + P_c(N))}{2*N} * \frac{CW_{min}}{2} \tag{11-4}$$

where $P_c(N)$ is the proportion of collisions experienced for each packet successfully acknowledged at the MAC layer ($0 <= P_c(N) < 1$).

The authors in [Heu03] propose a simple expression for $P_c(N)$. Their derivation is based on the observation that a host attempting to transmit a frame will eventually experience a collision if the value of the chosen backoff interval corresponds to the residual backoff interval of at least one other host. Such an approximation holds if multiple successive collisions are negligible. So, we have [Heu03]

$$P_c(N) = 1 - (1 - 1/CW_{min})^{N-1}. \tag{11-5}$$

Finally, the proportion p of the useful throughput that can be obtained by a host depends on the number of active hosts and is given by

$$p(N) = t_{tr} / T(N) \tag{11-6}$$

Performance Anomaly of 802.11b [Heu03]

In 802.11b, the hosts are allowed to use different data rates ranging from 1, 2, 5.5, or 11 Mbps. If we consider two classes of hosts: (a) (N-1) hosts transmitting at a high data rate, R, with an overall transmission time of T_f, and (b) one host transmitting at a low data rate, r, with an overall transmission time of T_s. Then,

$$T_f = t_{ov}^R + s/R + t_{cont} \tag{11-7}$$

where s is the length of data frame length in bits. Similarly,

$$T_s = t_{ov}^r + s/r + t_{cont} \tag{11-8}$$

Since the long term (but not short-term [Kok00]) channel access probability of CSMA/CA is equal for all hosts, then the utilization of any of the "fast" hosts will be

$$U_f = \frac{T_f}{[(N-1)T_f + T_s + P_c(N) * t_{jam} * N]} \tag{11-9}$$

where t_{jam} is the average time spent in collisions and

$$t_{jam} = \frac{2}{N} * T_s + (1 - \frac{2}{N})T_f \tag{11-10}$$

$$t_{jam} = P_s * T_s + (1 - P_s) T_f$$

where P_s is the probability of having a packet sent at the lower rate r involved in the collision. P_s can be computed as the ratio between the number of host pairs that contain the slow host and the total number of pairs that can be formed in the set of all hosts (assuming that all nodes are actively communicating), thus $P_s = (N-1)/[N*(N-1)/2] = 2/N$.

The throughput at the MAC layer of each of the (N-1) fast hosts will be

$$X_f = U_f * p_f(N) * R \tag{11-11}$$
where $p_f(N) = s/(R * T_f)$.

Similarly, the channel utilization of the slow host can be expressed as

$$U_s = \frac{T_s}{[(N-1)T_f + T_s + P_c(N) * t_{jam} * N]} \tag{11-12}$$

The throughput at the MAC layer of the slow host will be
$$X_s = U_s * p_s(N) * r \tag{11-13}$$
where $p_s(N) = s/(r * T_s)$.

So, we see that the fast hosts and the slow host will get the same throughput. Interested readers can refer to [Heu03] for plots of measured throughputs of TCP/UDP traffics with 3 hosts running at 11 Mbps and one host running at either 1, 2, 5.5, or 11 Mbps. For small number of hosts, their suggested approximation in Eqn. (5) gives throughput values that are close to measured values. Their work basically shows that the throughput of 802.11b WLAN can be much smaller than the nominal bit rate. The fair access mechanism provided by CSMA/CA causes a slow host transmitting at lower rate to capture the channel longer than those hosts that are emitting at the highest rate. This degrades the overall performance perceived by the users in the connected cell.

11.3.2 Understanding TCP Fairness in WLAN

One would expect that 802.11 MAC is designed to give all stations equal share of the WLAN bandwidth. However, via experiments, some researchers [Pil03] observe that the stations do not get equal share of the WLAN bandwidth. For example, with two or three mobile stations communicating to a server through a base station, the authors in [Pil03] observed that for the simple case of one mobile sender (upstream flow) and one mobile receiver (downstream flow), the sender receives 1.44 times the receiver's bandwidth. The authors also observed that such unfairness between upstream/downstream throughput becomes even more severe in the presence of background UDP traffic. There are many factors that impact the throughput ratio, namely base station buffer size, wireless link interference, implementation details of the 802.11 MAC etc. Thus, the authors in [Pil03] resort to simulations to try to understand this unfairness phenomena.

Via simulations, the authors did a couple of experiments. In the first experiment, the authors measure how the upstream vs downstream throughput ratio varies with the buffer size at the base station. The authors notice four regions of operations. The first region corresponds to the case where the buffer size is over 84 packets and the throughput ratio is one. This is the case where the buffer is large enough to accommodate the maximum receiver window of both flows, thus resulting in loss-free transmission in both upstream and downstream directions. The second region is when the base station buffer size is between 42 and 84 packets. In this region, the ratio between the uplink/downlink throughput decreases sharply from 10 to 1. In the third region (corresponding to buffer size between 6 and 42 packets), the ratio vary between 9 and 12. The fourth region is when the buffer size is smaller than 6 packets. The results for this region are very noisy.

To understand the issues behind the observed unfair behavior of TCP over wireless LAN, the authors started analyzing a simple case with one upstream and one downstream flow. When the window size is large, a loss of an acknowledgment packet has no real influence on the sender window size due to the cumulative acknowledgment nature of TCP. Thus, the upstream TCP window size will increase until it reaches the TCP receive window size, w, and remain at that size throughout the duration of the connection. The downstream TCP window size, however, changes considerably depending on the buffer size, B, and w since TCP reacts to loss of each data packet by halving its window. Let α be the number of ACK packets per data packet and assume that in steady state, the base station buffer has at most αw ACK packets and hence *(B- αw)* slots for downlink flow. Due to the TCP behavior in the congestion avoidance region, the downstream TCP window size will vary between *(B-αw)/2* and *(B- αw)*. Hence the ratio between the downstream throughput and upstream throughput is given by

$$R = \frac{w}{[\frac{3}{4} * (B - \alpha w)]} \qquad (11\text{-}14)$$

The authors also propose a more accurate model in [Pil03] using M/M/1/K queueing model to derive the ratio R. Interested readers can refer to [Pil03] to see the plots comparing the analytical and simulation results.

To solve this unfairness problem, the authors suggested using the advertised receiver window field in the acknowledgment packets toward the TCP sender. This field represents the available space at the receiver and lowering the receiver window can help throttle the TCP sender. If there are n flows in the system and the base station has a buffer size of B, then the receiver window of all the TCP flows is set to be the minimum of the advertised receiver window and. Their simulation results in [Pil03] indicate that this suggested approach can alleviate the observed unfairness problem.

11.4   Supporting Voice Services in 802.11b WLANs

With the popularity of WLAN data services, many people are interested in exploring how voice services can be provided in 802.11b WLANs. Papers presenting simulation results on using the polling in the PCF mode of IEEE 802.11 to support voice services include [Vis95],[Cro97], [Vee01]. Here, we summarize the analytical work presented in [Vee01].

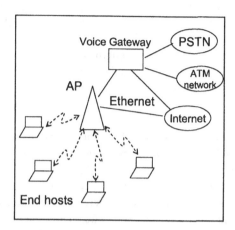

Figure 11-9. Network Architecture for Supporting Voice Services over Wireless LAN

In [Vee01], the assumed network architecture (see Figure 11-9) consists of an access point that has 2 network interfaces, an 802.11 interface and an Ethernet interface. The AP is connected to a voice gateway via wired connection. The mobile end hosts communicate with the access point via the 802.11 link. The voice gateway is connected to PSTN, Internet and possibly also ATM network. The voice gateway converts the 802.11 voice protocol stack to PCM voice for use on the PSTN, voice over an ATM Adaptation Layer (AAL) for ATM networks or voice over Real Time Transport Protocol (RTP)./UDP/IP for IP networks. For intra-AP calls and for calls between 802.11 users and PSTN/ATM/IP phones, the two ends (both the sending and receiving) are added to the polling list of an AP.

There are two possible modes of operation for transporting voice over 802.11. One can use a constant bit rate (CBR) mode in which calls are allocated their peak rate, or a Variable bit rate (VBR) mode where statistical multiplexing is used and silences in voice calls are used by other voice calls or data traffic.

Since the overhead of the MAC layer is high (34 bytes), the authors propose sending all the voice data generated within an interpoll period in one packet. In CBR mode, the AP will poll nodes once per superframe. With VBR, the AP can poll once per superframe, or multiple times per superframe. In CBR mode, the length of the Contention Period (CP) is determined by the number of admitted voice calls. In the VBR mode, when the AP completes polling all stations in the polling list, it sends CF-End and starts the CP, even if the CFPMaxDuration is not exhausted. However, it leaves the medium idle during the CP even if there are no data users.

For both modes, the superframe length is fixed because if the superframe length is varied, it will affect the delay variations of admitted calls. The number of voice calls that can be admitted is determined by the superframe length.

In the CBR mode, to determine how much time to allocate per call, we need to determine the maximum size of voice packets. Let the superframe time be T_{SF}. The largest voice packet size that can be created is $c(P_{min} + T_{SF})$ where P_{min} is the packetization delay to create one minimum size "sample" since we need to take care of the case where a voice packetization completes just after a poll and c is the voice codec rate. The time to handle a voice call in two directions is

$$T_w = \frac{(c*(P_{min}+T_{SF})+h+P)*4}{R} + 4*T_{sifs} \qquad (11\text{-}15)$$

where R is the transmission rate, h is the header overheads (RTP, LLC, MAC with WEP) in bits (57 bytes*8), and P is the physical layer header bits in bits.

To determine the maximum number of voice calls that can be supported, one needs to find the minimum duration of the *CP*, T_{CP} and then use T_{SF} minus this minimum duration for the CFP. T_{CP} is the sum of $T_{CP\text{-}min}$ and $T_{CP\text{-}Ext}$ where $T_{CP\text{-}min}$ includes the time to minimally send one frame as specified in [802.11] and $T_{CP\text{-}Ext}$ is the time needed for stretching CP to finish transmitting a maximum-size data packet. T_{max} is the time to send a maximum sized SDU that is fragmented into fragments of size f.

$$T_{CP} = T_{CP\text{-}min} + T_{CP\text{-}Ext} \qquad (11.16)$$

where

$$T_{CP\text{-}min} = 2\,T_{sifs} + 2\,T_{slot} + 8\,T_{ack} + T_{max} \qquad (11\text{-}17)$$
$$T_{CP\text{-}Ext} = T_{rts} + T_{sifs} + T_{cts} + T_{sifs} + T_{max} \qquad (11\text{-}18)$$
$$T_{max} = (m\text{-}1)\,([(f+h+P)/R]+ T_{ack} + 2\,T_{sifs}) + T_{last} \qquad (11\text{-}19)$$

where $m = \left\lceil \dfrac{S_{MaxSDU}}{f} \right\rceil$ and T_{last} is given by

$$T_{last} = (S_{MaxSDU} - f(m\text{-}1)+ h + P)/R + T_{ack} + 2\,T_{sifs} \qquad (11\text{-}20)$$

The maximum number of calls that can be admitted using CBR is

$$N_p = (T_{SF} - T_{CP} - T_{ovhd}) / T_v \qquad (11\text{-}21)$$

where

$$T_{ovhd} = (\frac{(B+P)}{R} + T_{sifs}) * \left\lfloor \frac{(T_{SF} - T_{CP})}{T_b} \right\rfloor + \frac{(CF_{End} + P)}{R} \qquad (11\text{-}22)$$

where B is the beacon size (in bits), R is the transmission rate, P is the physical layer header size, CF_{End} is the size of the message *CF-End*, T_b is the beacon interval.

For VBR mode of operation, the authors in [Vee01] computed the maximum number of calls, N_S, for two voice models, namely Brady's model [Bra69] and May-Zebo's model [Vee01]. Both models are ON-OFF Markov-Modulated Fluid models. They differ only in the mean holding times of the two states. For Brady's model, the mean ON period is 1 s, the mean OFF period is 1.35 s. For May-Zebo's model, the mean ON period is 0.352 s and the mean OFF period is 0.650 s. N_S is given by

$$\frac{1}{2pN_s} \sum_{k=2Np+1}^{2Ns} (k - 2Np) \binom{2Ns}{k} p^k (1-p)^{2Ns-k} \le \varepsilon \qquad (11\text{-}23)$$

where p is the probability that a sending end is active and ε is the statistical guarantee that the frame loss rate will not exceed this value.

Table 11-2 shows the maximum number of voice calls for B (Brady's model) and MZ (May-Zebo's model).

Table 11-2. Maximum Number of Supported Voice Calls

Tsf (ms)	DS (11 Mbps)		
	Np	Ns(B)	Ns(MZ)
75	22	41	51
90	27	52	65

Interested readers can refer to [Vee01] for the related voice delay analysis.

The above analysis shows that it is feasible to support voice calls using the contention-free period provided in the PCF access mechanism in 802.11b. However, there are some limitations with this access mechanism. Because the contention period may be stretched, the voice calls may have to endure some jitters. The Point Coordinator also needs to poll all stations on its polling list even if there is no traffic to be sent and hence the operation can be inefficient. In addition, current 802.11 standard also does not allow a polled station to transmit more than one Protocol Data Unit at any polled opportunity. Furthermore, the CFP repetition rate is not dynamically variable. One may want to perform some trade off between low latency applications requiring a fast repetition rate, and an efficient use of medium requiring slower repetition interval. Thus, it is obvious that better QoS features need to be incorporated into the existing 802.11 standard.

As multimedia applications become more popular in the Internet, mobile users also desire to enjoy such applications within WLAN. In July 1999, the 802.11 Working Group initiated a Study Group to define a new standard referred to as the 802.11e that provides QoS features in WLAN which we describe in the next section.

11.5   802.11e: Quality of Service in Wireless LAN

802.11e [802.11e] provides both reservation-based and prioritization services via a combination of some polling and random access techniques. Two new modes have been added to 802.11 in the new 802.11e standard, namely the Enhanced DCF (EDCF) and a Hybrid Coordination Function (HCF).

EDCF provides differentiated access to the medium. The 802.11e draft (D3.0) defines eight traffic categories for priority-based traffic. Each QoS-capable station (QSTA) marks their packets to indicate a specific premium service requirement. Nodes still contend for the medium but the channel access parameters e.g., the interframe spacing (AIFS), the contention window (*CWmin/max*) differ from one traffic class to another (refer to Figure 11-10.

EDCF provides relative QoS differentiation among traffic classes but do not provide any "QoS guarantees". The performance provided by this priority-based scheme is less predictable than a reservation-based method and may also suffer from network congestion. The QoS-capable Access Point (QAP) at most can only adjust the contention window and the Transmit Opportunity (TxOP) duration for each Traffic Class (TC).

The Hybrid Coordination Function (HCF) is an enhanced version of PCF that runs above EDCF using the PIFS to gain access to the medium. The Hybrid Coordinator (HC) is a type of Point Coordination (PC) that is

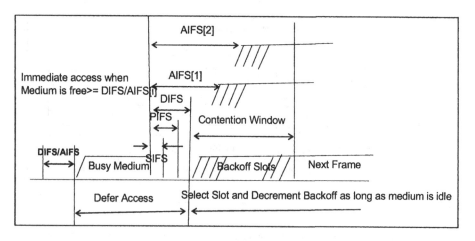

Figure 11-10. EDCF Mechanism

typically co-located with the QoS-enabled Access Point (QAP). The HC allocates the transmit opportunities to all QoS-enabled stations (QSTA) during both the CP and CFP periods, through centralized scheduling and taking into account the traffic contract and QoS requirements of each active connection. To give HC priority over the EDCF, AIFS must be longer than PIFS.

During a CP [Man02], each Transmit Opportunity (TxOP) begins either when the medium is determined to be available under the EDCF rules, i.e., after AIFS plus backoff time or when the station receives a special poll frame QoS CF-Poll from the HC. The QoS CF-Poll from the HC is sent after a PIFS idle period without any backoff. Therefore, the HC can issue polled TXOPs in the CP using its prioritized medium access. During the CFP, the starting time and the maximum duration of each TXOP is specified by the HC, again using the QOS CF-Poll frames. Stations will not attempt to get medium access on their own during the CFP, so only the HC can grant TXOPs by sending QOS CF-Poll frames. The CFP ends after the time announced in the beacon frame or by a CF-End frame sent by the HC. A typical 802.11e superframe is shown in Figure 11-11.

802.11e provides a controlled contention feature that allows HC to acquire updated information from the polled stations from time to time so that the HC can learn which station needs to be polled at which times and for what duration. The controlled contention mechanism [Man02] allows stations to request the allocation of polled TXOPs by sending resource

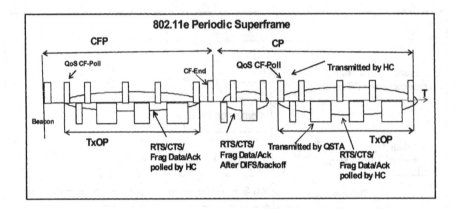

Figure 11-11. A Typical 802.11e Superframe [Man02] ©IEEE Copyright 2002

requests without contending with other EDCF traffic. Each instance of controlled contention occurs during the controlled contention interval which is started when the HC sends a specific control frame. This control frame forces legacy stations to set their NAV until the end of the controlled contention interval. The control frame defines a number of controlled contention opportunities (i.e., short intervals separated by SIFS) and a filtering mask containing the Traffic Classes(TC) in which resource requests may be placed. Each station with queued traffic for a TC matching the filtering mask selects one opportunity interval and sends a resource request frame containing the requested TC and TXOP duration, or the queue size of the requested TC. The HC provides for fast collision resolution by acknowledging the reception of a request via a control frame with a feedback field so that the requesting stations can detect collisions during controlled contention interval.

The authors in [Man02] have simulated a Quality of Service-enabled Basic Service Set (QBSS) with one access point and four stations. The access point implemented either the EDCF or HCF while the other four stations operate in EDCF. A radio channel error model as described in [Man01] is used. Transmission powers and distances between stations are chosen such that they are not hidden to each other with the selected PHY modes. All frames but the data frames are transmitted using 6 Mbps PHY mode. Data frames are transmitted using 24 Mbps PHY mode. Each station generates the same mix of offered traffic of three traffic classes: the low, medium, and high TC. For high-priority TC, MSDUs of 80 bytes arrive at constant periods. Each data stream from the medium- and low-priority TCs is 160 Kbps, and has a MSDU of 200 bytes with poisson inter-arrival times. The values of the EDCF parameters for the three TCs are shown in Table 11-3. The maximum achievable throughput for the three TCs presented in [Man02] are listed in Table 11-4.

Figure 11-12 is a throughput versus offered traffic plot for the three TCs. Each station has three data streams, one of each TC. The low-priority streams cannot carry their traffic for more than ten contending stations. The high priority streams always carry their traffic completely whereas the medium priority throughput decreases for 13 or more contending stations. Interested readers can refer to [Man02] for some simulation results on HCF performance. Additional 802.11e-based simulation studies can also be found in [Gar03],[Xia04].

Table 11-3. ED CF Parameters for the 3 TCs [Man02] ©IEEE Copyright 2002

	High	Medium	Low
AIFS	2	4	7
Cwmin	7	10	15
Cwmax	7	31	255
PF	2	2	2

Table 11-4. Maximum Achi evable Throughput for 3 TCs [Man02]©IEEE Copyright 2002

Frame Size	80 bytes	200 bytes	2304 bytes
High	3.5 Mbps	Not used	19.8 Mbps
Medium	Not used	6.22 Mbps	19.2 Mbps
Low	Not used	5.21 Mbps	18.2 Mbps

11.6   Other Related MACs

With the growing popularity of 802.11-based hotspot networks,e.g., at the airports, hotels, conference venues, etc 802.11 MAC has gained much attention. Many researchers have used this MAC for their performance studies in ad hoc network as well. The limited transmission range of 802.11b WLAN (a few hundred meters) means that each 802.11b WLAN serves only a small isolated area. On the contrary, 3G networks are considered wide-area networks that can support cell radii of up to 10 kilometers reliably. However, 3G networks are very expensive to deploy. As a result, there are strong incentives to use the 802.11b air-interface standard for outdoor purposes and possibly also with comparable cell radii. In the next subsection, we will summarize the work done in [Cla02],[Leu02] on outdoor 802.11-based Cellular Network.

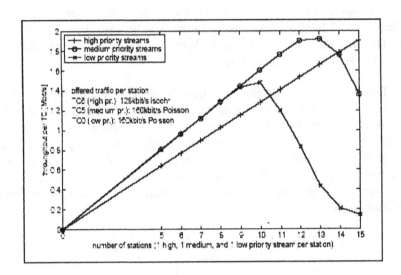

Figure 11-12. Thoughput vs Offered Traffic [Man02][21] ©IEEE Copyright 2002

11.6.1 Outdoor IEEE 802.11-Based Cellular Network

The authors in [Leu02] argued why it is infeasible to run PCF in outdoor IEEE 802.11-based cellular network. The most stringent requirement is that the ACK has to be received from the polled station to the AP within the SIFS interval, which is 10 µs for 802.11b networks. For an outdoor cellular network, the round trip signal propagation delay for a distance of 1.5 km requires 10 µs. Since at least several microseconds are needed for signal processing at the receiver, the link distance is likely to be limited to hundreds of meters as in the WLAN environments. However, for DCF operations, the authors argued that as long as the ACK is received before the ACK timer times out, the protocol can perform normally. The ACK timer out values can be as large as 50 µs (same as the DIFS interval). This means that one can allocate 10 µs for sender/receiver processing and 20 µs for propagation delay which translates to 6 km cell size. Thus, the authors propose enhancements to DCF for outdoor usage. Their enhancements extend the arrival delay for the ACK and CTS to the DIFS interval (in the order of 50 µs). The authors in [Leu02] provided some maximum MAC throughput results with different packet sizes and link distances. Interested readers can refer to [Leu02] for more detailed discussions of these results.

---

[21] The author wishes to thank Dr. S. Manfold for providing this curve and permission to use some texts from [Man02].

11.6.2   802.15.3 MAC

Apart from the 802.11 MAC, there is another emerging MAC for wireless personal area networks. The IEEE P802.15.3 High Rate (HR) Task Group (TG3) for Wireless Personal Area Networks (WPANs) is formed to define a new standard for high rate (20 Mbps or higher) WPANs. Besides a high data rate, the new standard is supposed to provide low-power, low-cost solutions to address the needs of portable consumer digital imaging and multimedia applications.

802.15.3 [802.15.3] is based on a centralized and connection-oriented ad hoc networking topology. At initialization, one node will assume the role of the coordinator of the WPAN (denoted as PNC). In addition to providing basic network synchronization timing, the PNC will also perform admission control, network resource allocation according to a pre-defined set of QoS policies and the amount of channel time (CT) resources available for data transfers, and perform power management (i.e. manages Power Save requests). The 802.15.3 time-slotted superframe structure comprises of three main sections: the beacon, the optional Contention Access Period (CAP), and the Contention Free Period (CFP) as shown in Figure 11-13.

The beacon is used to carry control information to the entire piconet (synchronization, Max Tx Power level), some specific application information, and the allocation of channel time (dedicated time slots) per stream for the upcoming superframe. The optional CAP is mainly used for the Authentication/Association Request & Response, CT Request/Response. The CFP is composed of Management Time Slots (MTS) and Guaranteed Time Slots (GTS), which are used for isochronous stream and asynchronous data connections. All transmission opportunities during the CFP begin at predefined times, relative to the actual beacon transmission time, and extend for predefined maximum durations, communicated in advance from the PNC to the respective devices using the traffic mapping information element conveyed by the beacon. During its scheduled GTS slot, a device may send an arbitrary number of data frames with the restriction that the aggregate duration of these transmissions does not exceed the scheduled duration limit. All GTS slots do not have the same level of persistence. Some are dynamic, meaning that their positions within the superframe may vary from one superframe to the next. Some others are pseudostatic meaning that the PNC may modify the position of these slots but it first needs to get the approval from both the Tx and Rx devices.

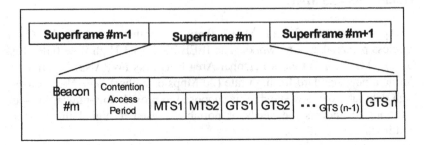

Figure 11-13. 802.15.3 Superframe Structure

Pseudostatic GTS slots are used to support CBR applications. The PNC is responsible for controlling the number of MTS slots that are allocated per superframe. Devices contend for the "Open & Association" MTS slots using the Slotted Aloha random access technique.

In 802.15.3, QoS is achieved through per flow reservation in a simple way: All QoS negotiations and flow control handling are supposed to be done at Layer 3. The network layer provides the link layer (SSCS+MAC) only with enough information to request the right level of resource at the air interface. The device is then supposed to query the PNC to discover if these QoS requirements can be granted by the CT Manager. If the request is accepted, a flow identifier will be established for that specific flow between the link layer and the IP layer. Such negotiation between the PNC and the requesting device significantly reduces the overall complexity, by not implementing intelligence in the lower layers. The reservation approach makes the best use of resources and gives more reliable QoS but may incur larger signaling overhead.

Acknowledgments

The authors wish to thank the authors in various papers and IEEE for their permission to use some of the figures in their papers.

11.7   Review Exercises

1. Evaluate the build out delay in Section 11.5.
2. Assume that there are two types of stations: real-time and non-real-time. Non-real-time stations use the following data backoff time.
    $DBT = rand(a,b)*Slot\_time$ where b grows exponentially for each retransmission attempt.

Real-time stations are given higher priority by providing a separate real-time contention window. The real-time backoff time (RBT) of a rt-station is defined as

$RBT = rand(c,d) * Slot\_time$ where $c < d < a$

The collided rt-station will retransmit the RTs in the following Slot_time with a probability p. With a probability $q = (1-p)$, it will defer and contend at the next real-time contention window.
Evaluate the throughput for both real-time stations and non-realtime stations. Interested readers can refer to [She01] for details of this analysis.

3. Compare the QoS approaches in 802.11e and 802.15.3. State the similarities and differences of the QoS approaches.

11.8 References

[802.11] IEEE 802.11 WG, Part II: Wireless LAN Medium Access Control (MAC) and Physical Layer (PHY) specifications, IEEE Standard, Aug. 1999.

[802.11a] IEEE 802.11 WG, Part II: Wireless LAN Medium Access Control (MAC) and Physical Layer (PHY) specifications: High-speed Physical Layer in the 5 Ghz Band, Supplement to IEEE 802.11 Standard, Sept 1999.

[802.11e] IEEE 802.11 WG, Part II: Wireless Medium Access Control (MAC) and Physical Layer Specifications: Medium Access Control Enhancements for Quality of Service, D4.4, June 2003.

[802.11g] IEEE standard P802.11g, Wireless Medium Access Control and Physical Layer Specification.

[802.15.3] IEEE 802.11 WG, Part 15: Wireless Medium Access Control and Physical Layer Specifications for High Rate Wireless Personal Area Networks (WPAN), D14, October 2002.

[Bia00] G. Bianchi, "Performance Analysis of the IEEE 802.11 Distributed Coordination Function," IEEE Journal on Selected Area on Communications, pp 535–547, March 2000.

[Bra69] P. Brady, "A Model for generating On-Off Speech Patterns in Two-Way Conversation," Bell Syst Tech Journal, Vol 48 No 7, pp 2245-2272, Sept 1969.

[Cal98] F. Cali, et al, "IEEE 802.11 Wireless LAN: Capacity Analysis and Protocol Enhancements," Proceedings of Infocom, 1998.

[Cro97] B. Crow, et al, "IEEE 802.11 Wireless Local Area Networks," IEEE Com. Magazine, Vol 35, No 9, pp 116-26, Sept, 1997.

[Cla02] M. Clark, et al, "Outdoor IEEE 802.11 Cellular Networks: Radio Link Performance," Proceedings of ICC 02.

[Des96] A. DeSimone etc, "Wireless data: systems, standards, services," J. of Wireless Networks, Vol 1, No 3, pp 241-254, Feb, 1996.

[Gar03] P. Garg, et al, "Using IEEE 802.11e MAC for QoS over Wireless," Proceedings of IPCCC, 2003.

[Heu03] M. Heusse, et al, "Performance Anomaly of 802.11b," Proceedings of Infocom 2003.

[Kok00] C. Koksal, et al, "An analysis of short-term fairness in wireless medium access protocols," Proceedings of ACM Sigmetrics, 2000.

[Leu02] K. Leung, et al, "Outdoor IEEE 802.11 Cellular Networks: MAC protocol design and performance," Proceedings of WCNC 02.

[Vee01] M. Veeraghavan, et al, "Support of voice services in IEEE 802.11 WLANs," Proceedings of Infocom, 2001.

[Man01] S. Mangold, et al, "An error model for radio transmissions of wireless LANs at 5 GHz," Proceedings of Aachen Symposium, Germany, 2001.

[Man02] S. Mangold, et al, "IEEE 802.11e Wireless LAN for Quality of Service," Proceedings of European Wireless Conference, 2002.

[May] C. May and T. Zebo, "A summary of speech statistics measured during the TASI-E-Rego Park-Ojus field trial," submitted for publication.

[Pil03] S. Pilosoft, et al, "Understanding TCP fairness over Wireless LAN," Proceedings of Infocom 2003.

[Qia02] D. Qiao, et al, "Energy-Efficient PCF Operation of IEEE 802.11a Wireless LAN," Proceedings of Infocom 2002.

[She01] S. Sheu, et al, "A Bandwidth Allocation/Sharing/Extension Protocol for Multimedia Over IEEE 802.11 Ad Hoc Wireless LANs," IEEE Journal on Selected Areas In Communications, Vol 19, No. 10, Oct, 2001.

[Vis95] M. Visser, et al, "Voice and data transmission over an 802.11 wireless network," Proceedings of 6th IEEE International Symposium on Personal, Indoor and Mobile Radio Communications (PIMRC) pp 648-52, Sept 1995.

[Xia04] Y. Xiao, et al, "Protection and Guarantee for Voice and Video Traffic in IEEE 802.11e Wireless LANs," Proceedings of Infocom, 2004.

Chapter 12

UPCOMING FEATURES FOR 3G NETWORKS

12. INTRODUCTION

With the advancement of various communication technologies, e.g., smart handheld devices and cellular networks with higher airlink bandwidth, there is an increasing demand for multimedia services at any place at any time. Multimedia services can be real-time or non-real-time. Video conferencing is an example of real-time multimedia service. An example of a non-real time multimedia service is multimedia messaging service (MMS), an enhancement of the popular SMS that allows multimedia messages. Multiple media types, e.g., video, audio, voice, and text are involved in multimedia services. Thus, multimedia applications impose new constraints to a network system. For example, the different media types within a session need to be synchronized before they can be presented simultaneously to the user. In addition, the time relation between successive data packets must be preserved for real-time services. The discussions in previous chapters indicate that the current 3G system is still far from an ideal system that can provide IP Multimedia Subsystem (IMS) service capabilities efficiently. Hence, the 3G system needs to be enhanced before flexible IMS services can be offered.

In addition, multicast communications for wireline users have been deployed in the internet for at least the past 10 years. With the widespread deployment of wireless networks, the fast-improving capabilities of mobile devices, and an increasingly sophisticated mobile work force, content and service providers are increasingly interested in supporting multicast communications over wireless networks. Many e-services can be made available if multicast services are available in wireless networks, e.g., mobile auctions, stock quote distributions, news, e-advertisements and e-coupons. The current 3G systems especially their airlinks are not designed to provide multicast services efficiently. Again, current 3G systems need to be enhanced to provide multicast/broadcast services.

Another popular service that is useful to small and medium size business is the one-way multiparty audio communication service, e.g., the push-to-talk service offered by Nextel Communications, Inc. A new generation of push-to-talk service over cellular, referred to as push to talk over cellular (PoC) service, has attracted much discussions in various industry standard bodies like Open Mobile Alliance (OMA), 3GPP and 3GPP2.

In this chapter, we will describe the ongoing work in 3GPP and 3GPP2 to design a beyond 3G system that can provide IMS capabilities, multicast services, and push-to-talk services efficiently. In addition, we also provide some brief discussions on how we envision the different access systems can be integrated together with a core network to provide seamless services to mobile users.

12.1   IP Multimedia Subsystem (IMS)

The third-generation networks such as UMTS and CDMA2000 are designed to provide multimedia services to mobile users. However, the end-to-end delay analysis in Chapter 9 indicates that the current Release 99 (or Release 4) UMTS network is far from perfect. The first-generation UMTS network is mostly designed to provide IP connectivity rather than to provide real-time IP multimedia services. In the initially deployed 3G network, wireless operators may offer non-standard enhanced IP services, e.g., I-mode service offered by NTT Docomo. However, it is difficult for the operators to offer services that will work across network boundaries.

There are advantages and disadvantages of providing only basic IP connectivity in 3G networks. Some advantages from the subscriber perspective [Won03] include: (1) the subscribers have the freedom to choose applications and service providers for their IP multimedia services, and (2) they can reuse many of their existing IP multimedia applications. However, some subscribers prefer to have the operators offer bundled services and pay only one bill. In addition, most of the existing IP multimedia applications, e.g., Yahoo messenger are designed for wired services and hence their performance may degrade in a wireless network when the last-hop communication link is subjected to fading degradation and hence not reliable. Even though in the past, network operators not familiar with business applications offered only IP connectivity to mobile users, such network operators are now keen to offer unique applications to mobile users to increase their revenues. NTT Docomo's successful I-mode service is a good example.

Thus, the 3G network needs to be enhanced to provide additional IP multimedia subsystem (IMS) capabilities. Currently, there is active work in 3GPP2 and 3GPP in enhancing 3G networks to provide real-time IP multimedia services efficiently. The designed IMS system can offer two levels of services [Won03]. The first level includes basic services such as multimedia session initiation, modification, termination, session forwarding etc. The second level of services include services that are provided by third-party providers using the capabilities provided by the network, e.g., location-based advertising, stock quote, or local news updates. This is made possible by the introduction of Open Service Access (OSA). OSA is a middleware framework that abstracts the network capabilities so that third-party applications can access the network capabilities in a secure manner without knowing how these capabilities are provided.

Figure 12-1 is a simplified picture of the UMTS Release 5/6 architecture. It shows the essential network elements used in providing real-time IP multimedia services. Only packet switched domain elements are shown since IMS uses only the packet-switched domain for transport and mobility management. A complete network architecture for Release 5/6 can be found in [23.002].

In Release 6 architecture, IMS signaling and session traffic are carried over IP networks. In addition, the Release 6 standard mandates that only IPv6 be used in the IMS domain [23.221]. Release 6's deployment is expected to be delayed because (1) there is a lack of large IPv6 deployment experience and (2) the IPv4 and IPv6 interworking issues need to be resolved before Release 6 architecture can be widely deployed.

The new entities introduced into the Release 6 architecture include Call Serving Control Function (CSCF), Media Gateway (MGW), Media Gateway Control Function (MGCF), Breakout Gateway Control Function (BGCF), Signaling Gateway (SGW), Home Subscriber Server (HSS), Media Resource Function Processor (MRFP), Media Resource Function Controller (MRFC), etc. We elaborate on the functions of these different new entities below:

- Call Serving Control Function (CSCF)

   The CSCF performs call control. There are different types of CSCFs: a proxy CSCF (P-CSCF) that is a CSCF used by a mobile station in its visiting domain; a serving CSCF (S-CSCF) that does session control; and an interrogating CSCF (I-CSCF) that is the entry point in a home network for all IMS-related signaling of its mobile

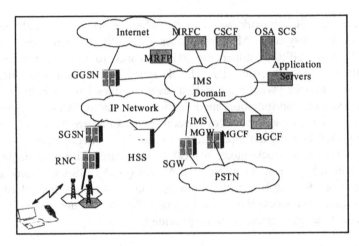

Figure 12-1 A Simplified Release 6 UMTS Network Architecture with IMS Capability

stations. SIP is used for signaling between the mobile station and CSCF, between CSCFs, between CSCF and MGCF and between CSCF and application servers.

- IMS Media Gateway (MGW), Media Gateway Control Function (MGCF), Signaling Gateway (SGW), and Breakout Gateway Control Function (BGCF)

Four new functional elements – the IMS MGW, the MGCF, the SGW, and the BGCF, are introduced to support the call scenarios between an IMS subscriber and a PSTN subscriber. The IMS MGW performs media translation. Apart from communicating with the S-CSCF, the MGCF controls the IMS media gateway and provides application level signaling translation between SIP-based and ISUP-based (used on PSTN side) signaling. The SGW translates the transport level signaling translation between IP-based and SS7-based transport. The identification of the network and the MGCF within that network where the breakout to the PSTN should occur is carried out by the BGCF.

- Home Subscriber Server (HSS)

The HSS is a master database that stores information such as subscriber profile and mobile location information. If there are many HSSs in a network, CSCFs will determine the HSS that has the desired subscriber profile during registration or session setup.

- Media Resource Function Processor (MRFP), Media Resource Function Controller (MRFC)

The media resource function processor (MRFP) and the media resource function controller (MRFC) support multiparty multimedia conferencing and media resource capabilities. The MRFC controls the media stream resources of the MRFP, which processes and mixes the actual media streams.

Application servers that provide features such as voice announcement, prepaid billing enforcement etc, are supported. There are two types of application servers: SIP-based and OSA-based. When OSA application server is used, an OSA service capability server is inserted between the OSA application server and the S-CSCF.

To obtain IP multimedia services using IMS, a mobile station needs to authenticate itself, and attaches to the UMTS network using GPRS attach procedures in the same manner in which it obtains IP connectivity service over GPRS. Then, the mobile activates a primary PDP context for signaling purposes. Next, the mobile will search for a P-CSCF in the network it is currently at so that it can perform IMS registration procedure. IMS registration is an application-level registration with the home network, during which the home network dynamically assigns an S-CSCF. The S-CSCF performs session control and keeps the P-CSCF information for this mobile during its registration lifetime.

Registration

The complete registration procedure is as shown in Figure 12-2 [23.228][Won03]. The mobile sends an SIP register that contains its home network domain name to the P-CSCF. The P-CSCF uses the home domain name to determine the entry point to the home network (i.e., I-CSCF) and forwards the REGISTER message to the I-CSCF. The I-CSCF queries the HSS to request information on S-CSCF capabilities, and the HSS responds with a list of available S-CSCFs. Based on the response from HSS and the required S-CSCF capabilities, the I-CSCF selects a S_CSCF for the mobile. The I-CSCF forwards the REGISTER message to the S-CSCF. The S-CSCF then interacts with the HSS to download relevant information from the subscriber profile as well as store the S-CSCF name at the HSS. Then, the S-CSCF returns a SIP 200 OK message to the mobile station via I-CSCF and P-CSCF.

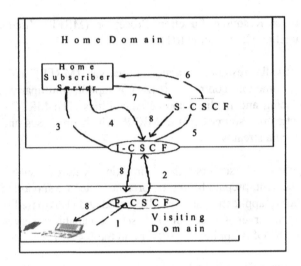

Figure 12-2 Registration Procedure

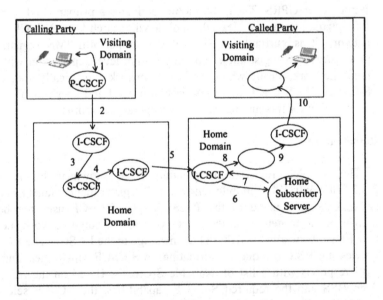

Figure 12-3 Call Flow for Wireless to Wireless Session Setup

Session Setup

A typical call flow for a wireless to wireless session setup is as shown in Figure 12-3 [23.228].

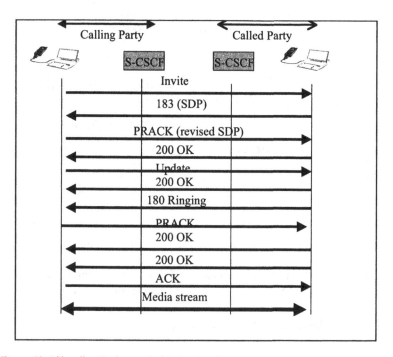

Figure 12-4 Signaling Exchange During Session Setup [Won03] ©IEEE Copyright 2003

The signaling exchanges during session setup are shown in Figure 12-4 [Won03]. The first step is for mobile 1 to send mobile 2 a SIP invite request message containing supported multimedia capabilities for preconditions negotiations. Mobile 2 returns the subset of multimedia capabilities that it can accept in a provisional 183 response. Mobile 1 then sends Mobile 2 a PRACK message to acknowledge receipt of the provisional response which includes the mutually agreed session parameters. After the resources for the session has been secured, mobile 1 sends an UPDATE message indicating success of the resource reservation. Negotiation of preconditions is completed when the 200 OK response is sent. Next comes the alerting phase, where the 180 Ringing response is sent. PRACK is used to ensure the reliable delivery of 180 Ringing. When the user has answered the session, a final 200 OK and ACK completes the session setup. Then, the media flow begins between the two mobiles. Note that the media flow does not flow through the CSCFs. If the called party is in the PSTN, then the S-CSCF in the caller's network will determine that the call should go to the PSTN, and will forward the INVITE to a BGCF in the same network. The BGCF will forward the INVITE to an appropriate MGCF. If the calling party is in the

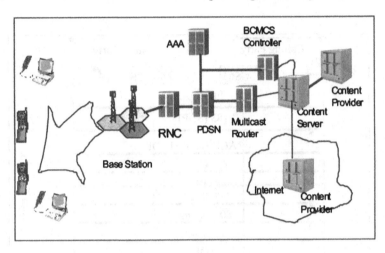

Figure 12-5 3GPP2 BCMCS Architecture [Wan04] ©IEEE Copyright 2004

PSTN, the MGCF in the IMS acts as the SIP end point that initiates requests on behalf of the PSTN, as if the signaling came from a S-CSCF.

12.2   Multicast/Broadcast Services

Recently, work has begun in both 3GPP and 3GPP2 to enhance 3G networks to support multimedia multicast/broadcast services. The goal is to design a system that can deliver multimedia multicast/broadcast traffic with minimum resource usage at both the radio access network and core network. In addition, the users that subscribe to multicast services expect minimum join/leave latencies and that the multimedia streams will be delivered without disruption as the mobile users move around. In this subsection, we briefly describe the current stage of the proposals discussed within 3GPP2 and 3GPP standard bodies.

12.2.1 Multicast/Broadcast Design for CDMA2000

Figure 12-5 illustrates a high-level view of the BCMCS architecture proposed in 3GPP2 [Wan04],[Bcmcs]. The multicast content originates from the BCMCS Content Provider which is either owned by the cellular operators or by a third party. For third-party contents, the subscribers either subscribe directly to the third party and hence need to establish security associations with the third party or the cellular operators negotiate for bulk access rate with the third party. In this case, security associations are negotiated between the content provider and the cellular service provider, if necessary, so that the subscribers from a particular cellular service provider can access the multicast content. For accessing those contents that the cellular operators own, a user merely needs to subscribe to the home cellular service provider so that a security association is established between the subscriber and the home cellular service provider. The BCMCS content server, owned by the cellular service provider, manipulates, and reformats the content received from the different content providers and sends them (possibly after encryption) to the CDMA RAN through the Packet Data Serving Node (PDSN). The PDSN is connected to the BCMCS Content Server via a multicast router (MR). If security is required, the PDSN can also be connected directly via an IPSEC tunnel to the BCMCS Content Server. The BCMCS Controller manages BCMCS sessions and provides session-related information to the PDSN, the mobile, and the content server (CS). In addition, it performs the authorization using the BCMCS user profile received from the Home AAA (Authentication, Authorization, Accounting) Server. The BCMCS Controller also performs security functions such as distributing security keys to the mobile users.

12.2.1.1 BCMCS Basic Operations

The cellular service provider or the content provider indicates the availability of the BCMCS to the users via the BCMCS service announcement and discovery [Wan04]. The service announcement allows the network to inform the users about the available services. In addition, the BCMCS service discovery mechanism also allows the users to request information about available services from the network.

Upon discovering the services, the Mobile Station (MS) user who desires to use the service needs to subscribe to the service provider by calling the operator or performs on-line subscription etc. During initial subscriptions, a security association needs to be established between the MS and the network. For example, a Registration Key (RK) is provisioned in the AAA subscription database and the User Identification Module (UIM) via the Over The Air (OTA) provisioning. After the RK is provisioned, the subscriber who receives other BCMCS programs can subscribe to them at any time.

On initial BCMCS subscriptions, the user can use its mobile device to acquire the BCMCS information via some system information acquisition procedures. The BCMCS session-related information includes the mapping between a BCMCF_FlowID and an IP multicast address/transport port number pair, and the broadcast access key from which session keys can be derived to decrypt the multicast data streams.

After acquiring such BCMCS information, the MS can determine whether a particular multicast service is available in a particular sector by interpreting the CDMA2000 overhead message broadcasted by a base station. If the MS user cannot find such information, and the overhead broadcast message from the base station indicates that BCMCS registration is permitted, then the MS user can activate the multicast service via the BCMCS registration request mechanism. If the registration is successful and the BCMCS bearer path is not established yet, then the first registered user for the service will trigger the PDSN to join the multicast group associated with that BCMCS_FlowID and consequently establish the bearer path from the PDSN to the RAN.

When the system determines that there are no more MSs listening to a particular multicast flow, the system may release the bearer path to save system resources.

Most of the enhancements required for supporting broadcast/multicast services occur at the RAN, e.g., the airlink and the signaling enhancements. For unicast service, the MS continues to monitor the paging channel for system information and pages for unicast services. When the MS subscribes to both unicast/multicast services, the MS may have to monitor more than one paging channels. A user needs to understand which forward supplemental channel (F-SCH) it needs to monitor to obtain sector-specific BCMCS information. The system needs to know which F-SCH a mobile is monitoring so that pages for the unicast services can be delivered uninterruptedly. A user in active state may be interested in continuing to monitor for BCMCS information. Thus, it is desirable to provide BCMCS flow to MSs on the traffic channel. Interested readers can refer to [Wan04] for more details on the signaling design proposed in 3GPP2 for supporting multicast/broadcast services.

12.2.1.2 Enhanced Physical Layer Design for CDMA2000

It is proposed that BCMCS be provided over the Forward Supplementary Channel (F-SCH) and the Forward Fundamental Channel (F-FCH). When the MS is idle, it does not have a connected reverse link to the base station. As a result, the system can support many such MSs without reaching the reverse link capacity limit. However, without a connected reverse link, the base station (BS) cannot have closed-loop power control for the forward link. It also implies longer handoff delay when MSs move between BSs. This is because the MSs need to go through the random access based Reverse Access Channel (R-ACH) or the Enhanced Access Channel (R-EACH) to report the need for handoff.

Since there is no power control for F-SCH, Reed-Solomon outer coding is proposed in [Aga04] to enhance transmission power efficiency for multicast services. The proposed outer coding operation is illustrated in Figure 12-6. Again, interested readers can refer to [Aga04] for more details on the CDMA2000 airlink design for broadcast/multicast services. The proposed CDMA2000 airlink design for broadcast/multicast services assumes a worst-case scenario, i.e., users at the edge of the coverage area where high speed broadcast service dictate the performance of the system.

A network-level simulation was performed in [Aga04]. The simulation was run in a network with 19 3-sector cells. The network was assumed to have a frequency reuse of 1 and soft-combining of broadcast signals from multiple sectors was allowed. Pedestrian A/B (3 km/hr) and vehicular A/B (120 km/hr) path loss models and channel impulse response models as specified in [Radio] were used. An interference-limited system was simulated. Time-varying log-normal shadowing with standard deviation of 10 dB, decorrelation length of 20 m was used. Two types of access terminals are simulated: (1) single antenna access terminal that can track eight paths and combine four paths in its rake receiver, and (2) dual antenna terminal capable of tracking 4 paths and combining four paths per antenna. Simulation results are presented in [Aga04] as the percentage of coverage area that can be served using a certain physical data rate and assuming a packet error rate (PER) requirement of 0.01. The PER is defined as the error rate of data carrying physical layer packets after the R-S deconding. The PER is computed assuming that even when more than K packets are erased from an error control block of N packets, the non-erased data packets are still forwarded to the higher layer.

With single-receiver access terminals, a data rate of 204.8 Kbps can be supported using a (32, 28) Reed-Solomon code in more than 90% of the coverage area (refer to Figure 12-7) assuming reuse factor of one, time-varying shadowing and soft-combining. With dual-receiver access terminals, a data rate of 409.6 Kbps can be supported using a (16, 14) Reed-Solomon code (refer to Figure 12-8).

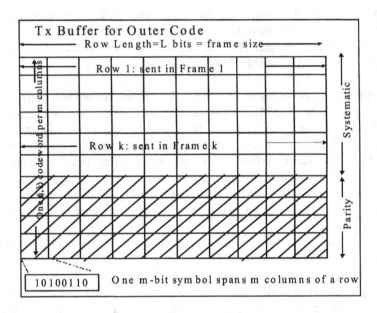

Figure 12-6 Transmit Buffer for Outer Coding [Aga04] ©IEEE Copyright 2004

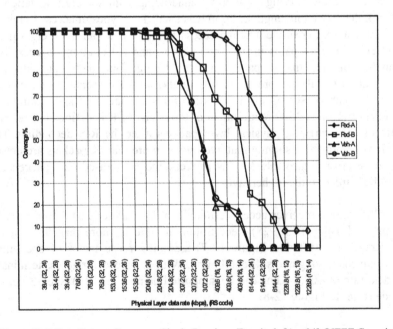

Figure 12-7 Data Rate vs Coverage: Single-Receiver Terminals [Aga04] ©IEEE Copyright
2004

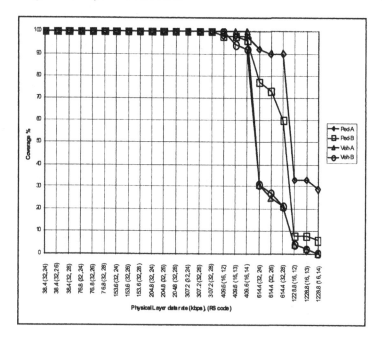

Figure 12-8 Data Rate vs Coverage: Dual-Receiver Terminals [Aga04] ©IEEE Copyright 2004

12.2.2   Multicast/Broadcast Design for UMTS

Figure 12-9 shows the 3GPP reference architecture for Multimedia Broadcast Multicast Services (MBMS) [23.246]. In the 3GPP reference architecture, a Broadcast Multicast Service Controller (BMSC) is available that plays the same role as the BCMCS controller in 3GPP2 system. The MBMS data source can be located within the BMSC or in a separate box. In this architecture, SGSN performs control functions for individual users and consolidates all individual users of the same MBMS service into one single MBMS service. One possible architecture option is to mandate that only GGSN needs to support IP multicast capability. Then, there is only one MBMS-GTP tunnel between each SGSN and GGSN for each single MBMS service. GGSN terminates the MBMS GTP tunnels and links such tunnels via IP multicast with the multicast data sources. Alternatively, if SGSNs also understand IP multicast, then GGSN can send multicast data using special IP multicast address (just for the UMTS network) and all SGSNs that need to receive this data can then subscribe to this special IP multicast address.

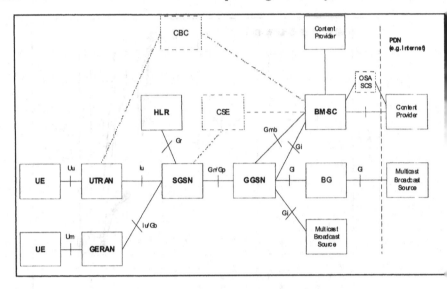

Figure 12-9 3GPP Reference Architecture for MBMS [23.246]

12.2.2.1 Required Architectural Enhancements for Supporting Broadcast/Multicast Services

One of the requirements for MBMS services is to minimize the MBMS service interruption when users move. An ideal all-IP based wireless network may mandate every base station to run IP multicast routing protocol. So, the mobile station only needs to perform the mobility update procedure to inform the new base station of the current IP multicast flows that it has subscribed to. The new base station can determine whether or not it is currently delivering such multicast contents over the air. If not, it will exchange signaling messages with a next hop access router to request for the required multicast traffic streams. Such signaling will stop at the first cross-over router (an access router or access gateway that already subscribes to the requested multicast streams). However, this approach requires many changes to current UMTS network since Release 99 base stations are provisioned to communicate with only one RNC and each RNC is configured to communicate with only one SGSN.

In the current 3GPP architecture, normally inter-SGSN handover is not performed until the user has terminated its current unicast data session. When a user moves to an RNC that is not controlled by the SGSN that the user has started with, then the new RNC will become the Drift-RNC

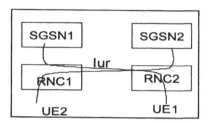

Figure 12-10 Multiple Iur Instances for MBMS Service

(DRNC) and the existing RNC becomes the Serving RNC (SRNC) and an Iur tunnel is built between the two RNCs so that the user can continue to receive unicast traffic. Let us see what needs to be enhanced to support multicast services. Assume that there are two User Equipment (UE) (refer to Figure 12-10).

UE1(2) is getting MBMS service from SGSN1(2) and it is moving towards SGSN2(1) respectively [Nor]. If we want to keep MBMS service control in the original SGSN, then, UE1 will still be served by SGSN1 in the control plane and an Iur interface needs to be created, similarly for UE2. To avoid a large number of Iur instances, it will be beneficial to have a separate control and user plane so that UE1 can get the data from SGSN2 rather than SGSN1. A common Iu user plane that is managed independently can be created at DRNC (refer to Figure 12-11). In Figure 12-11 (a), we show a UE1 that has both an active voice and MBMS session that is originally served by SGSN1 moving towards the area served by RNC2 and SGSN2. An Iu instance exists in SGSN2 to deliver an MBMS stream to RNC2 and RNC2 is currently delivering this MBMS stream using point to multipoint mode. After UE1 moves to the coverage area of RNC2 (shown in Figure 12-11 (b)), an Iur instance will be created to transport voice packets to RNC2 so that the two softhandoff legs can be created for the voice connection. The control connection for the MBMS session can still be controlled by SGSN1 but UE1 can obtain its MBMS data stream directly from the existing MBMS multicast stream from RNC2.

MBMS notification should be sent such that a UE in either IDLE or Radio Resource Connected (RRC) mode can receive it. In addition, the notification should be such that a UE will consume minimal power. One possibility is to create a new MBMS paging channel for MBMS notification such that after a UE checks the current paging channels whenever it wakes up, it will also check this new MBMS paging channel. In addition, a MBMS control channel (MCCH) will be used to provide the parameters of the radio

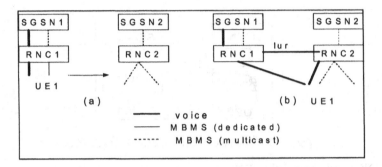

Figure 12-11 Switching Between P2P and P2MP Link for MBMS Service

bearers which carry the multicast data stream. The information in the MBMS notification message should be sufficient to let the UE determine if it needs to read the MBMS control channel (MCCH) for the information of the current radio bearers that are used to deliver the MBMS data that it is interested in.

12.2.2.2 UMTS Airlink Enhancement for Multicast Service

In the interference-limited CDMA system, the downlink capacity is limited by the base station transmission power. The point-to-multipoint communication nature of MBMS requires higher base station transmission power than the unicast service for the same application. The naïve reuse of the physical channel structure for the unicast service proves to be infeasible for the MBMS. Two main issues that must be addressed in order to achieve higher MBMS transmission efficiency are:

1. Lower target block error rate requirement than the unicast service for the same application
2. Coverage over all MBMS group members

Because of the intrinsic complexity associated with the multipoint-to-point feedback, the MBMS service has to use the unacknowledged transmission mode, which means no retransmission or ARQ is allowed to recover the lost data blocks. The only available error control scheme is through channel coding. The channel without retransmission is much less tolerant to the block errors than the channel with retransmission. Therefore, the MBMS service must have a lower block error rate target than the unicast service for the same application. This translates into higher target signal-to-interference ratio (SIR) requirement and higher transmission power.

There are several proposals for UMTS airlink enhancements to support broadcast/multicast services [Chu04]. In UMTS, two types of common transport channels [25.211] are defined to support point-to-multipoint communications, namely the Forward Access Channel (FACH) and the Downlink Shared Channel (DSCH). Dedicated Channel (DCH) is normally used to deliver unicast data service. When the number of MBMS users in a cell is small, it may be possible to deliver multicast traffic using dedicated DCH for each MBMS user. The advantage of DCH over FACH is that DCH is inner-loop and outer-loop power controlled. The disadvantages of DCH is that the radio resource of each channel is dedicated to each user. The transmission power and spreading codes cannot be shared by multiple users even if the data carried on each of the dedicated channels are identical. It is inefficient to use DCH to deliver MBMS data for a large MBMS group size.

Some alternative airlink enhancements that have been proposed are [Chu04]:

- Option 1: Using FACH

    From the network protocol perspective, the use of FACH to carry MBMS traffic is straightforward. It does not require any associated uplink channel. The MBMS group member simply needs to tune to the FACH channel to receive the MBMS data traffic. The main drawback of Option 1 is that FACH does not have an uplink channel to report the channel quality, and hence power control is not possible. If the MBMS group members' SIR is not known at the time when the MBMS service starts, the transmission power of the FACH has to be set such that it can cover the whole cell. For example, to cover one of the three 120-degree clover-leaf sectors of a cell, a 64-Kbps MBMS service carried on FACH will need an Ec/Ior of –0.6 dB or 87% of the total sector transmission power using the 3GPP Case 2 channel assumptions. Thus, a more power efficient method needs to be devised. One possibility is to use dynamic power setting approach described in [Chu04] where each MBMS user reports its path loss periodically and the base station changes its power setting according to the 95% worst case user's requirement.

- Option 2: Using DSCH

    DSCH is a closed-loop power controlled channel. The DSCH combines the merits of both FACH and DCH. However, to enable power control, each MBMS user needs to be assigned its own DPCCH uplink and downlink channels. The associated DPCCH overhead increases as the number of MBMS users increase. In addition, a clever method to update the transmission power level other than the obvious "power control based on the worst case user" method needs to be designed. Some preliminary results of the comparison between using DSCH and FACH with dynamic power settings are reported in [Wei02]. It shows that when the number of multicast receivers is 10, with 50% probability the DSCH solution consumes 1 dB more power than that consumed using the FACH with dynamic power setting solution.

- Option 3: Using Longer TTI and Space Diversity

In recent MBMS standard meetings, Single Transmitter Transmit Diversity (STTD) has been proposed to enhance the performance of FACH channel for delivering 64-Kbps MBMS traffic [Qua02]. The MBMS traffic is assumed to be delay tolerant. Hence, a longer TTI, e.g., 80 ms, has been proposed to carry the MBMS traffic. It has been shown that increasing the TTI from 20 ms to 80 ms without increasing the coding rate can provide 2- 4 dB gain [Qua02] in a typical urban pedestrian fading environment. However, the use of longer TTI introduces more complexity and larger memory space requirement in the mobile station.

- Option 4: Using Rate Splitting [Wei02b, Wei02c]

If the MBMS data stream is scalable, it can be split into several streams with different importance. Only the most important stream will be delivered to all the users in the cell to provide the basic MBMS service. The less important or enhanced streams can be sent at smaller power for the users with better channel conditions. Thus, the overall power requirement for delivering the MBMS service can be reduced.

- Option 5: Using Multiple DCHs and FACH [Wei02d]

In this approach, one can dynamically decide on a value K such that K users will be served using a broadcast FACH channel with *(M-K)* users being served using a point-to-point channels such that the total power required to support such a combination approach will be less than just using a broadcast FACH channel to reach all M users. The users that are in poorer radio conditions will be served using the point-to-point channels.

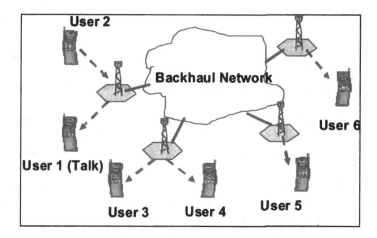

Figure 12-12 PoC Service Example: A One-to-Many Voice Group Session

All the above five options assume that only minor changes to existing airlink structures can be made to minimize the development cost for vendors. Interested readers can refer to [Chu04] for comparisons of these different options based on some simulation studies done by Bell Lab researchers. One can also explore the possibility of using High Speed Downlink Packet Access (HSDPA) channels which can achieve higher bit rates for delivering MBMS traffic. A preliminary discussion of this approach is described in [Ji02].

12.3 Push-to-Talk Over Cellular (PoC)

Push to Talk over Cellular (PoC) service (see Figure 12-12) allows users to engage in immediate communication with one or more users. PoC service is similar to a "walkie-talkie" application where a user presses a button to talk with an individual user or broadcast to a group of participants. The receiving participants hear the sender's voice either without any action on their part, or only after they accept the call. The communication is half-duplex, i.e., at most one person can talk at a time and all other parties hear the speech. This contrasts with voice calls which are full duplex. Both 3GPP and 3GPP2 have agreed to let Open Mobile Alliance (OMA) come up with the PoC requirements and design protocols to support PoC over 3G and beyond 3G networks.

The PoC service may support the following features:

- 1-to-1 communication feature. The voice communication attempt may either be accepted automatically or manually answered by the called subscriber
- 1-to-many communication feature. This feature enables a subscriber to set up a voice communication with a multiple number of other subscribers in which only one participant can speak at any time.
- Personal alert feature. This feature enables a subscriber to alert another subscriber. The alert expresses the calling subscribers' wish to communicate and to request the called subscriber to "call back."

The initial attempts were to provide only voice services but in the near future, other media types may be included. The PoC system should be designed such that users are allowed to use the PoC services only upon subscriptions. The subscribers need to be notified of ongoing PoC groups upon service activations and also be able to receive notifications to invited PoC groups.

A generic PoC architecture is shown in Figure 12-13 [PoCArch]. The radio access network and additional nodes that provide IP connectivity and IP mobility are required to support the PoC service. The PoC client, residing on the mobile terminal, is used to access PoC service. The PoC Server implements the application level network functions for the PoC service e.g. PoC session handling, SIP session handling, policy handling, participants' information dissemination, and generation of charging reports. The Group and List Management Server (GLMS) manages groups and lists. It provides list management operations to create, modify, retrieve and delete groups. The PoC Media Function implements media functionality e.g. media distribution, floor control functionality, collection of media quality information, etc. The Presence Server provides availability information.

It has been proposed that SIP and its enhancements are used as the signaling protocol [PoCSig] between a PoC client and the SIP/IP Core and between the SIP/IP Core and a PoC Server, between a PoC Client and a PoC Server as well as between a PoC client and a PoC Media Function. However, some engineers think that the text-based SIP protocol is not efficient for floor control update purposes and hence a lightweight protocol should be designed for floor control purpose. Some performance work needs to be done to quantify the benefit of having a separate lightweight protocol for floor control.

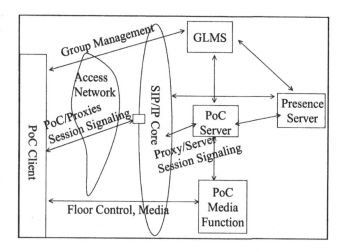

Figure 12-13 Proposed OMA PoC Architecture

Preliminary discussions on some enhancements to EGPRS/UMTS system to provide PoC services are summarized in another document [PoCUMTS]. For an example, it is suggested that an interactive class PDP context be activated to carry PoC Signaling and a streaming class PDP context be used to carry the RTP/RTCP packets.

12.3.1 An Example of SIP Call Flow for a PoC Session

In Figure 12-14, we give a simple SIP call flow for setting up a PoC session. A PoC client A initiates an adhoc PoC session or 1-to-1 PoC session by sending an INVITE request to the home PoC network. The SIP/IP core A routes the INVITE request to the PoC Server A based on the PoC address of the inviting PoC subscriber and PoC service indication. The PoC Server A takes the role of the PoC controlling and participating functions. The PoC Server A sends the invitation to the PoC clients of the invited PoC subscribers. When the first Ringing response is received, the PoC Server A forwards this ringing response towards the PoC Client A. Steps (5) and (6) are optional. The PoC Client A can send a SUBSCRIBE request in order to receive information about the result of the invitations. SIP/IP Core A will forward the SUBSCRIBE request to the PoC Server A. When the first PoC client accepts the PoC session request, the PoC Server A sends an OK response towards the PoC Client A. SIP/IP Core A forwards the OK response to the PoC Client A. Since one PoC participant is connected, the PoC Server A sends the floor control message to the PoC Client A. The PoC Client A then sends the media to the PoC Server A. When the final response

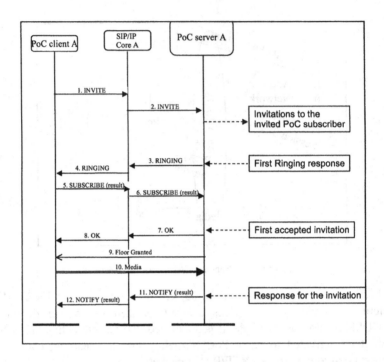

Figure 12-14. SIP Callflow for Setting Up an Ad hoc PoC Session

is received from an invited PoC subscriber, the PoC Server A sends the NOTIFY request to the PoC Client A. The SIP/IP Core A will forward the NOTIFY message to the PoC Client A.

Acknowledgments

The authors wish to thank many former Bell Laboratories colleagues for discussions on topics in this chapter, e.g., M. Dolan, T. Hiller, S. Palat, A. Lee, T. Ji, W. Luo, T. Hu etc. The work reported in Section 12.2.2.2 were done by Dr. W. Luo and Dr. M. Chuah with Dr. T. Hu while they were at Bell Laboratories. The author wishes to thank Dr. D. Wong for reviewing this chapter and providing useful comments.

The authors wish to thank 3GPP for giving permission to reproduce some diagrams from the 3GPP TSG standard documents. 3GPP TSs and TRs are the property of ARIB, ATIS, ETSI, CCSA, TTA, and TTC who jointly own the copyright for them. They are subjected to further modifications and are therefore provided to you "as is" for information purposes only. Further use is strictly prohibited.

12.4 Review Exercises

1. List the additional network entities that are added to UMTS network architecture to support IMS capability.
2. Describe the different types of CSCF and their functionalities.
3. Describe the similarity and differences between the 3GPP2 BCMCS architecture and 3GPP MBMS architecture.
4. Describe the enhancements made to the CDMA2000 airlink to support multicast services.
5. Describe the enhancements made to the UMTS airlink to support multicast services.
6. Describe the proposed OMA PoC architecture.

12.5 References

[23.002] 3GPP TS 23.002 V 6.1.0 (2003-03), "Network Reference Architecture (Release 6)."

[23.221] 3GPP TS 23.221 V5.4.0 (2002-03), "Architectural requirements (Release 5)."

[23.228] 3GPP TS 23.228 V3.2.0 (2002-03), "IP Multimedia Subsystem, Stage 2."

[23.246]3GPP TS23.246 V0.5.0 "MBMS Architecture and Functional Descriptions," 2003

[25.211] 3GPP TS 25.211 V3.11.0 (2002-06), "Physical channels and mapping of transport channels onto physical channels (FDD)," Release 99.

[Aga04] P. Agashe, et al, "CDMA2000 High Rate Broadcast Packet Data Air Interface Design," IEEE Comm Magazine, Feb, 2004.

[Chu04] M. Chuah, et al, "UMTS R99/R4 Airlink Enhancement for supporting MBMS services," Proceedings of ICC, 2004.

[Ji02] 3GPP TSG-RAN, WG#30, "HSDPA for MBMS," Lucent Technologies, Oct. 2002.

[Nor] 3GPP TSG-RAN WG3 #31, "MBMS – Iu and Iur Aspects," R3-021977, August 2002.

[PoCArch] Open Mobile Alliance, OMA-AD_PoC-V1_0-20031027-D, "Push to talk over Cellular (PoC)- Architecture."

[PoCSig] Open Mobile Alliance, OMA-SFD-PoC_V1_0-20030826, "PoC Signaling Flows."

[PoCUMTS] Open Mobile Alliance, OMA-User-Plane-V1.1.0-20030826," PoC User Plane: EGPRS/UMTS Specifications."

[Qua02] 3GPP TSG-RAN, WG1#28, "Evaluation of combining gains for MBMS (including STTD)," Qualcomm, R1-021234, October 2002.

[Radio] ITU-R.M.1225, "Guidelines for evaluation of radio transmission technologies for IMT-2000."

[Wan04] J. Wang et al, "Broadcast and Multicast Services in CDMA2000," IEEE Comm Magazine, February 2004.

[Wei02] 3GPP TSG_RAN, WG1#29, "Comparison of DSCH and FACH for MBMS," Lucent Technologies, R1-021325, November 2002.

[Wei02b] 3GPP TSG-RAN WG1#28, "MBMS Power Usage," R1-021234, October 2002.

[Wei02c] 3GPP TSG-RAN WG#29, "Power Control for FACH," Lucent Technologies, R1-021324, November 2002.

[Wei02d] 3GPP TSG-RAN WG1#28, "Power Usage for Mixed FACH and DCH for MBMS," Lucent Technologies, R1-021240, October 2002.

[Won03] K. Wong, V. Varma, "Supporting Real-Time IP Multimedia Services in UMTS," IEEE Communication Magazine, November 2003, pp 148–155.

Appendix

INTRODUCTION TO PROBABILITIES AND RANDOM PROCESS

A.1 The Basic Concept of probability

The theory of probability deals with averages of mass phenomena occurring sequentially or simultaneously: games of winning, telephone calls, check out service in stores and restaurants, birth and death events, electron emission, signal detection, system reliability and quality control, among many others.

There are several definitions of probability that have been playing important roles in the theory of probability. Among them, the axiomatic definition, the relative frequency definition, and the classical definition have been the basic and most used ones for centuries.

The axiomatic definition is relative recent and uses the concepts from set theory. According to the axiomatic definition, the probability $P(\mathcal{A})$ of an event $\mathcal{A}$ is a non-negative number assigned to this event:

$$P(\mathcal{A}) \geq 0 \tag{A-1}$$

The probability of the certain event equals 1:

$$P(\mathcal{F}) = 1 \tag{A-2}$$

If the event $\mathcal{A}$ and $\mathcal{B}$ are mutually exclusive, then

$$P(\mathcal{A} + \mathcal{B}) = P(\mathcal{A}) + P(\mathcal{B}) \tag{A-3}$$

This definition emphasizes the deductive character of a theory, avoids conceptual ambiguities, and provides a solid preparation for sophisticated applications. It is the best approach to introduce a probability and offers a very good starting point towards a deeper study of the probability theory.

The relative frequency definition is as follows. The probability $P(\mathcal{A})$ of an event $\mathcal{A}$ is the limit

$$P(\mathcal{A}) = \lim_{n \to \infty} \frac{n_{\mathcal{A}}}{n} \tag{A-4}$$

where n is the number of trials and $n_{\mathcal{A}}$ is the number of occurrences of $\mathcal{A}$.

This approach to probability is fundamental in the applications of probability without a priori definition. However its use as the basis of a deductive theory must be challenged.

The classical definition of probability is used to determine probabilistic data and as a working hypothesis. The probability $P(\mathcal{A})$ of an event $\mathcal{A}$ is determined a priori without actual experimentation. It is given by the ratio

$$P(\mathcal{A}) = \frac{N_{\mathcal{A}}}{N} \qquad\qquad\qquad (\text{A-5})$$

where N is the number of possible outcomes and $N_{\mathcal{A}}$ is the number of outcomes that are favorable to the event $\mathcal{A}$.

The classical approach to probability is a consequence of the principle of insufficient reason, which leads to the assumption that all outcomes of the event $\mathcal{A}$ are *equally likely*. This definition has an underlying conceptual ambiguity and can be questioned on several grounds. For example, the use of *equally likely* is made of the concept to be defined. The definition can be applied only to a limited class of problems.

One of the applications of probability theory is induction. In other words, from past observations of the probability $P(\mathcal{A})$ of an event $\mathcal{A}$ in a given experiment, how do we draw conclusions about the occurrence of this event in future performances of this experiment. The theory of probability provides us the analytical tools to answer this kind of questions subjectively and objectively.

A.2   Random variable and random process

A.2.1 The Concept of a Random Variable

We shall use the definition in [Pap91] to define a random variable. A random variable is a number $\mathbf{x}(\zeta)$ assigned to every outcome ζ of an experiment. It is a function whose domain is the set $\mathcal{F}$ of experimental outcomes. In other words, given an experiment specified by the space $\mathcal{F}$, the field of subsets of $\mathcal{F}$ is called events, and the probability assigned to these events is a function that is called the random variable. All random variables are written in boldface letters, as the conventional way of representation. The formal definition of a random variable is as follows.

A random variable **x** is a process of assigning a number $x(\zeta)$ to every outcome ζ. The resulting function must satisfy the following two conditions but is otherwise arbitrary:

1. The set $\{\mathbf{x} \le x\}$ is an event for every x.
2. The probabilities of the events $\{\mathbf{x} = \infty\}$ and $\{\mathbf{x} = -\infty\}$ equal 0.

A.2.2 Distribution and Density Function

The *cumulative distribution function* (CDF) of a random variable **x** is defined as:

$$F_x(x) = P\{\mathbf{x} \le x\} \tag{A-6}$$

for every x from $-\infty$ to ∞ .

Properties of distribution functions:

1. $F(+\infty) = 1, \; F(-\infty) = 0$ $\qquad\qquad$ (A-7)

2. If $x_1 < x_2$, then $F(x_1) \le F(x_2)$ $\qquad\qquad$ (A-8)

3. If $F(x_0) = 0$, then $F(x) = 0$ for every $x \le x_0$ $\qquad$ (A-9)

4. $P\{\mathbf{x} > x\} = 1 - F(x)$ $\qquad\qquad$ (A-10)

5. The function $F(x)$ is continuous from the right:

$$F(x^+) = F(x) \tag{A-11}$$

6. $P\{x_1 < \mathbf{x} \le x_2\} = F(x_2) - F(x_1)$ $\qquad\qquad$ (A-12)

7. $P\{\mathbf{x} = x\} = F(x) - F(x^-)$ $\qquad\qquad$ (A-13)

8. $P\{x_1 \leq x \leq x_2\} = F(x_2) - F(x_1^-)$ (A-14)

A.2.3 The Density Function

The derivative

$$f(x) = dF(x) / dx$$ (A-15)

of $F(x)$ is called the *density function* (or probability density function, i.e., PDF) of the random variable x.

Properties of density functions:

    1. $f(x) \geq 0$ (A-16)

    2. $P\{x_1 < x \leq x_2\} = \int_{x_1}^{x_2} f(x)dx$ (A-17)

A.2.4 Moments and Conditional Distributions

Mean and variance

The mean and variance of a random variable are important quantities in the study of random variables.

The *expected value* or *mean* of a random variable x is defined as

$$\mu_x = E\{x\} = \int_{-\infty}^{\infty} x f_x(x)\, dx$$ (A-18)

The *variance* of a random variable x is defined as

$$\sigma_x^2 = E\left\{(x - \mu_x)^2\right\} = \int_{-\infty}^{\infty} (x - \mu_x)^2 f_x(x)\, dx$$ (A-19)

where μ_x is the mean of x as defined in (A-18). The positive constant σ_x is called the *standard deviation* of x.

From the definition, we have

$$\sigma_x^2 = E\left\{(x - \mu_x)^2\right\} = E\left\{x^2\right\} - E^2\left\{x\right\} \tag{A-20}$$

Moments

Moments are often used quantities to characterize a random variable.

The *moment* of a random variable x is defined as

$$m_n = E\left\{x^n\right\} = \int_{-\infty}^{\infty} x^n f_x(x)\, dx \tag{A-21}$$

where n is an integer.

Obviously, the mean and variance of x are special cases of the moments with particular n following the equalities,

$$\mu = m_1 \qquad \sigma^2 = m_2 - m_1^2 \tag{A-22}$$

The constants μ_x and σ_x give only a limited characterization of the random variable and its density function. Other moments provide additional information that is quite useful. For some density function, knowledge of the mean and variance determines the distribution uniquely. However, most random variables may have the same mean and variance but with different density functions. Under certain conditions, if the moments for every n are known, the density function can be determined completely.

Conditional distributions

Conditional distributions are the distributions under a certain condition. They can be expressed as conditional probabilities. The corresponding density functions are obtained by appropriate differentiations.

For example, the conditional distribution $F_y(y \mid \mathbf{x} \le x)$ and the density function $f_y(y \mid \mathbf{x} \le x)$ are expressed as

$$F_y(y \mid \mathbf{x} \le x) = \frac{\Pr\{\mathbf{x} \le x, \mathbf{y} \le y\}}{\Pr\{\mathbf{x} \le x\}} = \frac{F(x, y)}{F_x(x)} \tag{A-23}$$

$$f_y(y \mid \mathbf{x} \le x) = \frac{\partial F(x, y)/\partial y}{F_x(x)} \tag{A-24}$$

The conditional density of y assuming $\mathbf{x} = x$ is of particular interest. It can be proved that

$$f_y(y \mid \mathbf{x} = x) = \frac{f(x, y)}{f_x(x)} \tag{A-25}$$

We have the following probabilities of the conditional density functions.

1. $f_y(y \mid x) = \dfrac{f(x, y)}{f_x(x)} \qquad f_x(x \mid y) = \dfrac{f(x, y)}{f_y(y)}$ \hfill (A-26)

2. If the random variables $\mathbf{x}$ and $\mathbf{y}$ are independent, we have

$$f(x, y) = f(x) f(y)$$
$$f_y(y \mid x) = f(y) \text{ and } f_x(x \mid y) = f(x) \tag{A-27}$$

3. $f(x, y) = f(y \mid x) f(x)$ \hfill (A-28)

Bayes' theorem

$$f(x \mid y) = \frac{f(y \mid x) f(x)}{f(y)} \tag{A-29}$$

Conditional expected values
The conditional mean and variance of $\mathbf{y}$ are

$$\mu_{y|x} = E\{y \mid x\} = \int_{-\infty}^{\infty} y f(y \mid x) \, dy \tag{A-30}$$

$$\sigma_{y|x}^2 = E\left\{ \left(y - \mu_{y|x} \right)^2 \mid x \right\} = \int_{-\infty}^{\infty} \left(y - \mu_{y|x} \right)^2 f(y \mid x) \, dy \tag{A-31}$$

12.5.1.1 FunctCalculation of the mean and variance
f random variables forms a new random variable with outcomes consisting of the events determined by the function itself.

Let us assume that x is a random variable and $y = g(x)$ is a function of x. Then y is defined as a new random variable with its domain including the range of the random variable x. The distribution function of y can be expressed by the distribution function of x and the function $g(x)$.

$$F_y(y) = P\{y \le y\} = P\{g(x) \le y\} \qquad \text{(A-32)}$$

From $y = g(x)$, we can find the values of x such that $g(x) \le y$.

For example, $y = ax + b$

If $a > 0$, then

$$F_y(y) = P\{ax + b \le y\} = P\left\{x \le \frac{y-b}{a}\right\} = F_x\left(\frac{y-b}{a}\right) \qquad \text{(A-33)}$$

If $a < 0$, then

$$F_y(y) = P\{ax + b \le y\} = P\left\{x \ge \frac{y-b}{a}\right\} = 1 - F_x\left(\frac{y-b}{a}\right) \qquad \text{(A-34)}$$

Calculation of the density function

We shall express the probability density function of $y = g(x)$ in terms of the density function of x and function $g(x)$. It can be proved that the relations between $f_y(y)$ and $f_x(x)$ given $y = g(x)$ is

$$f_y(y)\,dy = f_x(x)\,dx \qquad \text{(A-35)}$$

Thus

$$f_y(y) = f_x(x)/g'(x) \qquad \text{(A-36)}$$

where $g'(x)$ is the derivative of $g(x)$.

For example, $y = ax + b$, we have $g'(x) = a$. Thus

$$f_y(y) = \frac{1}{|a|} f_x(x) = \frac{1}{|a|} f_x\left(\frac{y-b}{a}\right) \qquad \text{(A-37)}$$

Calculation of the mean and variance

The expected value of the random variable y can be expressed directly in terms of the function $g(x)$ and the probability density function $f_x(x)$. From equation (A-32), we can show that

$$m_y = E\{y\} = E\{g(x)\} = \int_{-\infty}^{\infty} g(x)f_x(x)\,dx \qquad \text{(A-38)}$$

A.2.5 The Concept of a Random Process

A random process $x(t)$ is defined as a family of time functions assigned to every ζ outcome a function $x(t, \zeta)$. Equivalently, a random process is a function depending on parameters t and ζ. If the domain of t is a set of real axis, then $x(t)$ is a continuous-time process. If the domain of t is the set of integers, then $x(t)$ is a discrete-time process.

We can interpret $x(t)$ as an ensemble of functions $x(t, \zeta)$. For a fixed ζ, $x(t)$ is a time function with the given sample point of the process. In this case, t is a variable and ζ is fixed. For a specific t (or t is fixed and ζ is a variable), $x(t)$ is a random variable with distribution

$$F(x,t) = P\{x(t) \le x\} \qquad \text{(A-39)}$$

For multiple specific times t_1, t_2, ..., t_n, $x(t)$ is a set of random variables. The *nth-order* distribution of $x(t)$ is the joint distribution

$$F(x_1, x_2, \cdots, x_n; t_1, t_2, \cdots, t_n) = P\{x(t_1) \le x_1, x(t_2) \le x_2, \cdots, x(t_n) \le x_n\} \text{(A-40)}$$

The statistical properties of a real random process $x(t)$ are determined by its nth-order distribution as shown above.

For a complex process $z(t) = x(t) + jy(t)$, the statistical properties are determined by the joint distribution of the two real random processes $x(t)$ and $y(t)$. The joint distribution of $x(t)$ and $y(t)$ is represented by the distribution of the random variables $x(t_1)$, ..., $x(t_n)$, $y(t_1')$, ..., $y(t_n')$.

Properties of a random process

For many applications, the expected value of $x(t)$ and of $x^2(t)$ are often used. They are defined as follows:

Mean   The mean of $x(t)$ is the expected value of the random variable $x(t)$:

$$m(t) = E\left\{x(t)\right\} = \int_{-\infty}^{+\infty} x f(x,t) dx \qquad (A\text{-}41)$$

Autocorrelation   The autocorrelation of $x(t)$ is the expected value of the product $x(t_1)\,x(t_2)$:

$$R(t_1,t_2) = E\{x(t_1)x(t_2)\} = \int_{-\infty}^{+\infty}\int_{-\infty}^{+\infty} x_1 x_2 f(x_1,x_2;t_1,t_2) dx_1 dx_2$$

$$(A\text{-}42)$$

Covariance   The covariance of $x(t)$ is:

$$C(t_1,t_2) = E\left\{x(t_1)x(t_2)\right\} - E\left\{x(t_1)\right\}E\left\{x(t_2)\right\} = R(t_1,t_2) - m(t_1)m(t_2)$$

$$(A\text{-}43)$$

A.3 Common Distributions of Random Variables and Processes

The definition of random variables starts from known and conducted experiments. In the development of random variables and statistic theory, we often consider random variables or processes with specific distribution or probability density functions without referring to any particular experiments. In the following, we describe several special distributions of random variables and processes that are most commonly used in modern digital communications and wireless communication systems.

A.3.1 Normal or Gaussian Distribution

A random variable x is *normal* or *Gaussian* distributed if its probability density function is represented by

$$f(x) = \frac{1}{\sigma\sqrt{2\pi}} e^{-(x-\mu)^2/2\sigma^2} \qquad (A\text{-}44)$$

where μ is the mean and σ is the standard deviation. The constants μ and σ determine the density function uniquely.

Gaussian distribution is widely used to characterize various random behaviors in the communication systems. For example, the random noise in the transmission media is often considered as Gaussian distributed. Another example is the sum of a large number of independent random variables approaches to a normal distribution. This is illustrated by the well-known *central limit theorem*, which is expressed in the following.

The Central Limit Theorem

Given $x = x_1 + x_2 + \cdots + x_n$ where x_i are independent random variables, the distribution of x approaches to a normal distribution when n increases. The mean and variance of x are $\mu = \mu_1 + \mu_2 + \cdots + \mu_n$, $\sigma^2 = \sigma_1^2 + \sigma_2^2 + \cdots + \sigma_n^2$.

The approximation in the central limit theorem and the required value of n for a specified error bound depend on the density function of each individual random variable x_i. In general, the value n = 30 is adequate for most applications if the random variables are i.i.d. (independently, identically distributed).

A.3.2 Log-Normal Distribution

A random variable x is log*normal* distributed if its probability density function is represented by

$$f(x) = \frac{1}{\sigma x \sqrt{2\pi}} e^{-(\ln x - \mu)^2 / 2\sigma^2} \tag{A-45}$$

Lognormal random variables are closely related to the normal random variables. Suppose that another random variable y is a function of random variable x. If $y = \ln x$ and x is lognormal distributed with density function as (A-45). It is easy to show that y is normal distributed with parameters μ and σ. It can be proved that the mean and variance of x and y have the following equalities [Sch92].

$$\mu_x = e^{\mu_y + \frac{1}{2}\sigma_y^2} \tag{A-46}$$

$$\sigma_x^2 = e^{\sigma_y^2 + 2\mu_y}\left(e^{\sigma_y^2} - 1\right) \tag{A-47}$$

The central limit theorem for products holds for the lognormal approximation. Given $x = x_1 x_2 \cdots x_n$ where x_i are independent *positive* random variables, the distribution of x approaches to a lognormal distribution when n increases. The mean and variance of x are given by

$$\mu = \sum_{i=1}^{n} E\left(\ln x_i\right) \quad \sigma^2 = \sum_{i=1}^{n} Var\left(\ln x_i\right) \tag{A-48}$$

This can be proved from the central limit theorem via taking a logarithmic operation on the variable.

A.3.3 Uniform Distribution

A random variable x is *uniform* distributed between x_1 and x_2 if its probability density function is represented by

$$f(x) = \begin{cases} \dfrac{1}{x_2 - x_1} & x_1 \leq x \leq x_2 \\ 0 & \text{otherwise} \end{cases} \tag{A-49}$$

A.3.4 Binomial Distribution

A random variable x has a binomial distribution of order n if it takes the values $0, 1, \ldots, n$ with probability

$$P\{x = k\} = \binom{n}{k} p^k q^{n-k} \quad p + q = 1, \ 0 \leq k \leq n \tag{A-50}$$

The probability density function of x is thus a sum of impulses as

$$f(x) = \sum_{k=0}^{n} \binom{n}{k} p^k q^{n-k} \delta(x - k) \tag{A-51}$$

A.3.5 Poisson Distribution

A random variable x is Poisson distributed if it takes the values $0, 1, \ldots, n, \ldots$ with probability

$$P\{x = k\} = e^{-a} \frac{a^k}{k!} \quad k = 0, 1, \ldots \tag{A-52}$$

where a is a constant parameter.

The probability density function of x is thus a sum of impulses as

$$f(x) = e^{-a} \sum_{k=0}^{\infty} \frac{a^k}{k!} \delta(x - k) \tag{A-53}$$

The moments of x are expressed as

$$m_n = E\{x^n\} = e^{-a} \sum_{k=0}^{\infty} k^n \frac{a^k}{k!} \tag{A-54}$$

The mean and variance of a Poisson distributed random variable equal to the parameter a:

$$\mu = m_1 = a \qquad \sigma^2 = m_2 - m_1^2 = a^2 + a - a^2 = a \tag{A-55}$$

The probability expression in equation (2-53) is the outcome of a limit theorem, or the **Poisson Theorem** for random points.

Poisson Theorem

If $n \to \infty \quad p \to 0 \quad np \to a$

then

$$\frac{n!}{k!(n-k)!} p^k q^{n-k} \to e^{-a} \frac{a^k}{k!} \tag{A-56}$$

where $p + q = 1$.

An important application of the Poisson theorem is the approximate estimation of Poisson points. We place n points randomly in the interval $(0, T)$. We define a random variable n, with its value equals to the number of points in the interval $(0, t_0)$. Hence

$$P\{n=k\} = \binom{n}{k} p^k q^{n-k} \quad \text{where } p = \frac{t_0}{T}, \ q = 1-p \tag{A-57}$$

Assuming that $n \gg 1$, $t_0 \ll T$, and $\lambda = n/T$ remains a constant as n and T increase, we have

$$P\{n=k\} = e^{-\lambda t_0} \frac{(\lambda t_0)^k}{k!} \tag{A-58}$$

Therefore the number of points in an interval of length t_0 is Poisson distributed with parameter $a = \lambda t_0$ where λ is the number of points in a time unit, or the density of the points. λ is often called the arrival rate in a queuing system.

The random Poisson points experiment is fundamental in the probability theory and has wide applications. Telephone calls, queuing analysis, electron emission, short noise and many others, are illustrations of Poisson distribution.

A.3.6 Chi-Square Distribution

A random variable x is Chi-square distributed with n degrees and denoted by

$$f(x) = \frac{1}{2^{n/2} \Gamma(n/2)} x^{n/2-1} e^{-x/2} U(x) \tag{A-59}$$

where

$$\Gamma(n/2) = \int_0^\infty x^{n/2-1} e^{-x} dx \quad n > 0 \tag{A-60}$$

A.3.7 Rayleigh Distribution

A random variable x is *Rayleigh* distributed if its probability density function is represented by

$$f(x) = \frac{x}{\alpha^2} e^{-x^2/2\alpha^2}, \ x \geq 0 \qquad\qquad \text{(A-61)}$$

Suppose a complex signal is the sum of two quadrate Gaussian signals as

$$z = re^{j\theta} = x + jy \qquad\qquad \text{(A-62)}$$

where x and y are i.i.d. Gaussian random variables with zero mean and variance of σ^2. The envelope of the signal $r = \sqrt{x^2 + y^2}$ is a random variable with density function $f(r)$. Since x and y are i.i.d., we have

$$f(x,y) = f(x)f(y) \qquad\qquad \text{(A-63)}$$

Since

$$\frac{\partial f(r)}{\partial x} = \frac{df(r)}{dr} \frac{\partial r}{\partial x} = \frac{df(r)}{dr} \frac{x}{r}$$

Taking differentiation on (A-63) with respect to x, we have

$$\frac{x}{r} f'(r) = f'(x) f(y) \qquad\qquad \text{(A-64)}$$

or

$$\frac{1}{r} f'(r) = \frac{1}{x} f'(x) f(y) \qquad\qquad \text{(A-65)}$$

Similarly, we have

$$\frac{1}{r} f'(r) = \frac{1}{y} f(x) f'(y) \qquad\qquad \text{(A-66)}$$

From (A-65) and (A-66), we show that the probability density function $f(r)$ is circular symmetric.

The mean and variance of a Rayleigh distributed random variable are given by

$$\eta_r = E\{r\} = \sqrt{\frac{\pi}{2}}\sigma \tag{A-67}$$

$$\sigma_r^2 = E\{r^2\} - E^2\{r\} = \left(2 - \frac{\pi}{2}\right)\sigma^2 \tag{A-68}$$

Rayleigh distribution is widely used to model the wireless fading channel [Rap][Jak74]. The received signal over the air usually does not have a direct line-of-sight to the transmitter. The transmitted signal is scattered by randomly placed obstructions and results signals with different attenuations and phases. This is so called multipath propagation. As a consequence, the received signal is the sum of the signals from multiple paths.

Each scattered wave is a complex signal with a random amplitude and a phase. Let a_i and θ_i denote the amplitude and phase of the *i*th wave respectively, the received signal from *n* waves is thus

$$S_{received} = re^{j\theta} = \sum_{i=1}^{n} a_i e^{j\theta_i} \tag{A-69}$$

Further, we have

$$S_{received} = re^{j\theta} = \sum_{i=1}^{n} a_i \cos(\theta_i) + j\sum_{i=1}^{n} a_i \sin(\theta_i) \triangleq x + jy \tag{A-70}$$

The amplitude of individual wave a_i is random, and the phase θ_i is uniformly distributed. Since the number of scattered wave *n* is large, from the central limit theorem, *x* and *y* approach to Gaussian distribution with zero mean and same variance. From (A-70), we conclude that the envelop of the received signal has a Rayleigh distribution as shown in equation (A-61).

For a special case *n=2* and $y = \sqrt{x}$ in the Chi-square distribution, we obtain the Rayleigh distribution with the density function denoted by (A-61).

A.3.8 Rician Distribution

A random variable x is *Rician* distributed if its probability density function is represented by

$$f(x) = \frac{x}{\sigma^2} e^{-\left(x^2+a^2\right)/2\sigma^2} I_0\left(\frac{ax}{\sigma^2}\right), \ x \geq 0 \qquad \text{(A-71)}$$

where $I_0(\cdot)$ is the modified 0th order Bessel function given by

$$I_0\left(\frac{ax}{\sigma^2}\right) = \frac{1}{2\pi} \int_0^{2\pi} e^{\frac{ax\cos\theta}{\sigma^2}} d\theta \qquad \text{(A-72)}$$

The Rician distribution is another widely used model for the wireless fading channels. It is used to model the fading channel when there is a direct line-of-sight path. The Rayleigh fading model characterizes the scenario of indirect multi-path propagation. However, in some environment such as the indoor propagation, there is a direct path dominating over the indirect multiple paths. Therefore the received signal is the sum of the scattered and direct signals, which is given by

$$S_{received} = re^{j\theta} \triangleq (x+a) + jy \qquad \text{(A-73)}$$

Following the same procedures in deriving (A-61), we can obtain the probability density function of the received envelop as in (A-71).

It should be noted that if $a=0$ in (A-73), we obtain the Rayleigh distribution. If a/σ is very large, the signal will be dominated by the in-phase part $(x+a)$. The distribution of r will approach to the distribution of x with a mean value of a.

A.4   Review Exercises

1. Consider a random variable $x = x_1 + x_2 + \cdots + x_n$. Suppose that random variables x_i are i.i.d. and uniformly distributed in the interval $(0,1)$. Express the density function $f_x(x)$ with $n=2$ and $n=5$. Compare the density function with the normal distribution using the central limit theorem.
2. Prove the Bayes' theorem in equation (A-29).

3. Prove that the mean and variance of a lognormal random variable are expressed as in equations (A-46) and (A-47).
4. Derive the Rician distribution of the envelop of the received signal as in equation (A-71).

A.5 References

[Pap91] Athanasios Papoulis, *Probability, Random Variables, and Stochastic Processes, Third Edition*, McGraw-Hill, 1991.

[Sch92] S. C. Schwartz, Y. S. Yeh, "On the Distribution Function and Moments of Power Sums With Log-Normal Components," *The Bell System Technical Journal, Vol. 61, No. 7, pp. 1441-1462*, September 1992.

[Rap] T. S. Rappaport, T. Rappaport, *Wireless Communications: Principles and Practice (2nd Edition)*, Prentice Hall, December 2001.

[Jak74] W. Jakes, *Mobile Communications*, CRC Press, 1974.

Index